Nuclear Power and
the Public Safety

Nuclear Power and the Public Safety

A Study in Regulation

Elizabeth S. Rolph

Lexington Books
D.C. Heath and Company
Lexington, Massachusetts
Toronto

Library of Congress Cataloging in Publication Data

Rolph, Elizabeth.
  Nuclear power and the public safety.

  Bibliography: p.
  Includes index.
  1. Atomic power—Law and legislation—United States. 2. Atomic power-
plants—Safety regulations—United States. I. Title.
KF2138.R64   343'.73'092   78-24794
ISBN 0-669-02822-3

Published simultaneously in Canada

Printed in the United States of America

International Standard Book Number: 0-669-02822-3

Library of Congress Catalog Card Number: 78-24794

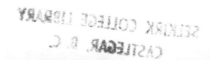

To my parents
Elizabeth Kales and Francis Howe Straus

Contents

List of Figures

List of Tables

Preface

This study of the Atomic Energy Commission's regulation of commercial nuclear power plants grew out of an earlier study I did jointly with several other Rand colleagues on the development and commercialization of the light water reactor.[a] At that time I was primarily interested in understanding how the regulation of nuclear power reactors had evolved and what effect that regulation had on commercial development. However, as I concluded that work, the material began to suggest a new set of very interesting issues.

It seemed to me that nuclear power was an example *par excellance* of a special class of technology that was commanding increasing public and regulatory attention. The substances and technologies in this class have several characteristics in common. They are reasonably new to the marketplace. They are expected to enjoy broad use. There are substantial uncertainties about the consequences of large-scale use of the substance or technology. And most often the uncertainties cannot be resolved easily, since experience with the product is limited and lab-scale tests leave many questions unanswered. In short, regulation of this class requires weighing generally known, immediate benefits against generally unknown, deferred, and *possibly* very large hazards. Examples that fit into the group include poisonous substances, often with mutagenic or carcinogenic properties (for example, benzene, radioactive material, certain pesticides, asbestos, and a host of others too numerous to name) and substances or technologies that threaten to disrupt the ecological balance in a fundamental way (for example, use of fluorocarbons, DDT, nuclear power, and others).

Because of its special characteristics, this group of substances and technologies could be expected to impose a special stress on the regulatory process for a number of reasons. First, the stakes are clearly very high, and therefore people are apt to feel strongly and push hard on the political system. Second, there is considerable evidence in the psychology literature that suggests individuals show strongly biased preferences when uncertainties are great or probabilities low but stakes high. And third, given the original uncertainties, demands for regulatory flexibility and change would seem inevitable.

The fact that the stakes are high and the institutions are being pushed, in some sense, to their limits makes regulation of this class of products an especially attractive vehicle for gaining insights into the regulatory process. And once I had the data on the AEC in hand, using it to address a set of more general regulatory questions seemed only sensible.

---

[a]Reports published under the NSF grant include: *Nuclear Reactors for Generating Electricity: U.S. Development from 1946 to 1963,* by Wendy Allen, R-2116-NSF, June 1977; *Electric Utility Decisionmaking and the Nuclear Option,* by Arturo Gandara, R-2148-NSF, June 1977; *Development and Commercialization of the Light Water Reactor, 1946-1976,* by Robert Perry and others, R-2180-NSF, June 1977; and *Regulation of Nuclear Power: The Case of the Light Water Reactor,* by Elizabeth Rolph, R-2104-NSF, June 1977.

Broadly speaking, I wanted to explore what institutional mechanisms and decision algorithms provided the regulatory base for dealing with this group of products. I wanted to learn to what degree decisions were based on technical judgments and to what degree political and economic considerations influenced outcomes. Furthermore, I hoped that a close examination of a particular regulatory experience might suggest a set of criteria that could be used to judge the quality of regulatory institutions, their structure, and their performance.

Obviously this study of the AEC experience is only a first step. Before we can know whether that history is unique or is generalizable in its fundamental characteristics, comparisons will have to be drawn between regulation of similar and dissimilar products and between a variety of regulatory authorities. But my hope is that this is at least a start. This start should provide interesting and provocative reading for the political scientist who may intend to extend the base of analysis; for the policy analyst and decision-maker who may use this work as the basis for more immediate and practical decisions; or for the interested citizen who may want insight into the issues of the nuclear debate. While it is perhaps impossible to satisfy such a diverse group, the story of AEC regulation is a dramatic one and the questions it raises are important to us all. So if any such history is of broad interest, this one should be.

I am deeply indebted to a number of people and organizations for their support and assistance. Foremost among them is the National Science Foundation, sponsor for the initial research. I am also grateful to the Rand Corporation for supporting the additional research and rewriting necessary to turn the original report into a book.

While many individuals contributed mightily to the thinking and writing of this book, I especially want to thank Bill Ahern, Rae Archibald, Carl Builder, Arturo Gandara, and Dick Hewlett, all of whom labored through the original NSF report, offering valuable insights and suggestions. I would also like to thank Bill Mooz for so generously sharing his cost analysis data and results, and George Eads, Phyllis Ellickson, and Adele Palmer for listening to ideas, reading drafts, and most importantly for planting the seeds for many new and useful ideas. Finally, I would like to pay handsome tribute to Robert Perry and to John Rolph, for without their intellectual and moral support there would have been no book.

Topanga, California   *Elizabeth S. Rolph*
January, 1979

1 Introduction

Despite the repeated warnings of the Office of the Surgeon General, people still smoke like chimneys. But they don't like to see their children smoke and the thing they like least of all is to see chimneys smoke.[1]

—Sherman R. Knapp, President
Connecticut Yankee Atomic Power Co.

In 1954 Congress passed a new Atomic Energy Act mandating the Atomic Energy Commission to develop nuclear power for commercial uses and at the same time to insure that its use imposed no undue risks. The Commission's new dual mission captured the dilemma the federal government would soon be facing on a broad variety of products: how to insure the public safety without stifling economic growth and progress.

In accordance with the Reserved Powers provision of the Constitution, providing for the public health and safety was historically a matter left to the states. But by 1906, growing abuses in food and drug marketing combined with increasing public awareness of these abuses to force passage of the federal Pure Food and Drugs Act. This act, holding the federal government responsible for the safety of food, drugs, cosmetics, and medical devices, opened the way for an expanding federal role in protecting the public health and safety. With the exceptions of major amendments to the act passed in 1937 and the Federal Insecticide, Fungicide, and Rodenticide Act of 1947, Congress chose not to extend federal regulatory controls for the next fifty years.

In recent years, however, the regulatory environment has shifted markedly. The rapidity of technological evolution is, in many instances, outstripping our ability to assess the risks of application. The public has a vastly heightened appreciation of "risks," both everyday and novel, and seems generally more inclined to expect government to intervene as its protector. And Congress, responding to these changes, has significantly expanded federal regulatory authority.

Obstacles to Regulation

Within the broad range of its present regulatory responsibilities, government is faced with a subset of products that poses some particularly obstinate problems. This subset includes both new technologies or processes and newly exploited

1

substances enjoying rapid development and application on a large scale. The products have in common the fact that their large-scale application *may* incur harmful consequences that are far-reaching and irreversible, but the hazards associated with their use have not yet been well documented. While the regulatory problems posed by these products may not be unique, as a group they are formidable.

Of the four principal regulatory obstacles, *uncertainty* is foremost. Because the technology or substance is new and untried, regulators cannot judge how likely it is to be "misused." That is to say, how likely are accidents? What is the probability that plutonium might be diverted for terrorist use? Are growers likely to use available toxic pesticides on table crops, contrary to regulations?

Moreover, the consequences of large-scale application are uncertain for several reasons. Again, because the technology or substance is new, there are major experimental and theoretical gaps in our understanding of its impact on man and his habitat. We have accumulated little operating or use experience. Also, in recent years we have increased our understanding of delicate ecological interrelationships and have vastly improved our ability to measure the presence of potentially hazardous substances. This, in turn, allows us to infer *potential* dangers in countless new quarters. But, at the same time, we have not been able to devise and conduct experiments that would conclusively establish the existence of hazards from such measurable low-level insults.

A dramatic example of this change can be found in early DDT studies. In a series of tests on the toxicity of DDT conducted by the Agricultural Research Administration in 1944, researchers wanted to know if ingested DDT was passed into the blood. Since reliable methods for measuring the chemical had not yet been developed, cows were fed large doses, then their blood was fed to flies. When 50 percent of the flies fed on DDT-dosed blood died, as compared with less than 2 percent for the control group, it was presumed DDT was passed from the digestive tract to the bloodstream.[2] Today we accurately measure the presence of such chemicals in parts per billion or even trillion. But we rarely can measure the effects of these small concentrations of toxins.

The level of perceived uncertainty is increased because the public shows a diminishing willingness to rely on traditional sources of expertise for safety assessments and policy judgments. This means that staff recommendations from regulatory bodies such as the Atomic Energy Commission, the Environmental Protection Agency, or the Food and Drug Administration or assessments or recommendations from independent scientists or scientific organizations do not go unchallenged and cannot automatically provide the basis for regulatory decisions. Moreover, the challenges generally mean scientific opinion is pitted against opposing scientific opinion either in public hearings or in the media. Thus uncertainties are highlighted.

A second obstacle arises from *distributional differences.* The risks and the benefits of using a new technology or substance are most frequently distributed

over different populations. For instance, nuclear power plants produce electricity primarily for use in major urban areas, yet they must be located in more
remote, rural areas. And it is the nearby rural residents and population centers
that bear the greater risk. This kind of disparity makes equity hard to define or
achieve.

A third obstacle, conceptually related to the second, stems from the fact
that different segments of society attach *different values* to economic growth
and to freedom from risk, and trade-offs between the two inevitably must be
made. The regulatory objective has been to make the product or its use "safe."
Yet "safe," as traditionally and colloquially used, means "freed from harm or
risk" or "secure from threat of danger...."[3] Clearly *total* freedom from risk or
security from danger cannot be guaranteed with the broad use of even the most
benign of technologies or chemicals. Nothing is risk-free.

Trade-offs must be made, but regulatory authorities find themselves hard-
pressed to make them. Generally the new products or technologies are the result
of very large private research, development, and marketing investments. Therefore, producers and users are inclined to exert considerable political pressure
opposing regulatory controls that increase costs, delay the introduction of a
product, or constrain its use. Conversely, the present public attitude toward risk,
or at least toward new sources of risk, appears to be growing more conservative.
It is also possible that this apparent change is owing more to an increasing public
awareness of potential risks than to actual changes in the public's willingness to
accept risk. Whichever is the root cause, the result is the same: growing public
skepticism and efforts to constrain use of processes and substances that present a
potential large-scale threat.

A final often misunderstood point is that regulation is, of necessity, a
political process. Precisely because there are uncertainties and distributional
differences and differences in values, and because equity is a key attribute of a
viable regulatory decision, the role of expertise is limited.

A useful way of defining "safe," one presently gaining currency, is as that
level of *risk* judged *acceptable*. In this context, *risk* is defined as the probability
that harm will occur at all multiplied by the severity of the consequences if it
does occur. Thus *risk* objectively measures the potential hazard, while *safety*
reflects a subjective judgment of the acceptability of that hazard.[4] Risk is
legitimately the subject of scientific investigation. Biologists or physicians are
best equipped to provide information on the likely carcinogenic effects of
exposure to specified chemicals. Engineers, seismologists, and statisticians may
be able to calculate the likelihood that a dam will withstand specified kinds of
earthquakes and then the likelihood of that kind of quake in the first place.
Scientists, however, cannot determine when something is *safe* or *safe enough,*
because that is a matter of preference or judgment. Does the group want to live
with the risks described by the scientist as accompanying the product, pay for
reducing the risks, or, instead, forego the product?

In recent years, social scientists, particularly economists, have sought to expand the purview of cost-benefit analysis to include preferences and nonquantifiable amenities. In theory it might be possible for experts using such an expanded analytical technique to estimate the acceptable level of risk for an individual or group. But the methodologies so far explored fall far short of being able either to produce a comprehensive list of relevant characteristics to be measured or to persuasively reflect the public values attached to those characteristics. Moreover, even if cost-benefit analysis or some variant thereof could assess costs and benefits, the fact they are not borne uniformly means their weighing remains the province of the political process.

Because regulatory decisions reflect political choices, in theory Congress should make them. But in fact, the responsibility has generally been delegated to regulatory commissions or executive agencies for two reasons. First, government and industry rarely admit that a choice between benefits and risks must be made. Instead, they tend to suggest that with proper expert advice, the risks can be controlled while a full measure of benefits continues to be enjoyed. And second, because the issues have often been extremely technical, it seems sensible to refer them to a more specialized body. On occasion, Congress has specified with some precision the standards a regulatory body must enforce. But most often delegated discretion is exceedingly broad.

The Regulatory Process: Measures of Adequacy

Because decisions must be made from day to day, regulators (both Congress and the Executive) have established procedures and safety criteria, but often without appreciating the extent of the uncertainties, the differences in distributional impact, the need for trade-offs between safety and economic growth, or the possibility of changes in the political environment. Failure to appreciate these factors has resulted, as all the participants are only too well aware, in a regulatory instability that threatens both industry's economic position and the public's confidence in government's ability to protect. Since the introduction of potentially hazardous technologies and substances appears to be on the increase, the need to regulate in cases of high uncertainty and substantial conflict over the values attached to risk and economic benefit seems also likely to increase. Therefore, it is important that we develop a systematic framework for understanding and evaluating regulation under these conditions.

There appear to be four measures that reflect the quality of the regulatory process. First, *the regulatory decisions should reflect accurate perceptions of the known risks and of the uncertainties that come with using the product or technology*. That is, the technical information upon which regulatory decisions are based must be objectively sound. And if there is a large degree of uncertainty about a major consideration, decisions should reflect that fact. Without a sound information base, regulators cannot control the risk so that it matches an "acceptable risk." and still more important, when the basis for past decisions is found wanting, future decisions are more likely to be suspect and to be challenged, thereby causing delays and confusion.

Second, *regulatory decisions should reflect substantial political sensitivity.* They need to balance or compensate for imbalances in the distribution of costs and benefits derived from use of the regulated product. They also need to mirror, at least roughly, a public consensus on the weights to be accorded economic growth and public safety.

If regulatory decisions do not reflect a sensitivity to distributional and value differences, given a reasonably democratic system of governance, public opposition to those decisions is sure to surface. And sooner or later, that opposition will disrupt the regulatory process and perhaps threaten the future of the regulatory authority.

As is true of so much in this era of rapid technological evolution and change, there is the potential for substantial instability in the regulatory system. Frequently, decisions are made and standards are adopted on the basis of very incomplete information on both the risks and on the technical feasibility and costs of reducing them. New information may challenge the wisdom of the decisions and create strong pressures to change them. Moreover, technologies and products evolve over time, often changing the risk-burden they impose on society. Technological change may then also create pressures to reconsider regulatory requirements. Last, but certainly not least in importance, is the potential for change in public values and public willingness to bear previously acceptable levels or types of risk. Change in this dimension almost certainly will lead to strong public pressure to amend regulatory standards. Then, because change is so likely, it is most important to explore how regulators incorporate new information, account for technological evolution, and accommodate public values. Regulatory decisions need to reflect all these changes even though the unevenness may work a temporary hardship upon the regulated industry.

Third, *decisions should be timely and, for the most part, predictable.* If they are not, delays and unproductive investment will inevitably alter the economics of the product and force suppliers and customers out of the marketplace for reasons that have nothing to do with safety. This is not to say the regulatory authority should not take the time it needs to accumulate and assess the necessary technical information and to assess the public's attitudes. Nor can it ignore changes in public attitudes in order to maintain the predictability of its decisions. But it should function as efficiently and predictably as it can, consistent with its obligations to protect the public. Inevitably, there will be constant tension between the authority's need to make rapid, predictable decisions and the need to resolve uncertainties and respond to constantly changing technologies and public attitudes.

Finally, *the public should have confidence in the impartiality and compe-*

tence of the regulatory body. Confidence leads the public to accept regulatory decisions without obstructing the process or demanding conclusive evidence of the authority's conservatism at every turn. It also denies any persistent challengers the broad political support necessary for meaningful opposition. Clearly, public confidence is, in large measure, dependent upon the regulatory body's meeting the above three criteria. For the most part, its decisions should be timely, be objectively sound, and reflect a sensitivity to distributional and value difference, or the public will eventually lose confidence in it. Then its authority is open to challenge. But public confidence is independent of the first three measures in that even when a regulatory authority meets the first three criteria, it may not enjoy public confidence. This might be true, for instance, when the authority loses public confidence after a period of poor performance, then reforms itself but is unable to win back public support.

Tools

The regulatory body has a variety of means by which to improve its performance. They fall into three main categories: procedures, information resources, and control over the pace of development and commercial use of the product.

Any regulatory body must adopt a host of procedures for reviewing information, setting standards, and applying those standards. Those procedures can serve a number of objectives. Most often, they determine the efficiency and equity of the process. Moreover, the use of certain procedural mechanisms—for instance, public hearings—can facilitate consensus formation or acquaint decision-makers with public attitudes.

A regulatory body must also have access to a broad range of technical information and advice. Agency-managed research programs, industry-supplied research, expert advisory panels, public investigative hearings, and independent research findings all qualify as information resources. And these resources can serve several functions. They can provide underpinnings for sound, objective regulatory decision-making. Certain of the more credible or reputable resources can also be used to prove to a skeptical public the validity of the regulatory body's assumptions and conclusions and to legitimize regulatory decisions.

Finally, a regulatory body can use its authority to pace the development and commercialization of a product it controls. It can curb the speed of deployment or the speed of technological change or both while waiting for information to be developed or a consensus on risk acceptability to form.

Understanding to what ends and how well a regulatory agency uses these tools, then, becomes the point of departure in assessing how well a regulatory agency is doing its job and in what areas it might be deficient.

The Case under Study

The above discussion of measures and tools establishes a general framework for examining and evaluating the regulation of technologies or products where the potential for harm is great and the uncertainties high. It is my intention, then, to apply this framework to the case of the Atomic Energy Commission's regulation of commercial nuclear power.

As a case study from which to generalize, the regulation of commercial nuclear power has its strength and its weaknesses. It can be reasonably argued that because Congress gave the Atomic Energy Commission the dual mandate of protecting the public health and safety *and* of developing and promoting nuclear power, the behavior and decisions of the Commission may differ from those of single purpose regulatory bodies. Therefore the Commission would be a poor case from which to generalize. At the same time, it can be argued that the dual mandate of the Atomic Energy Commission is analogous to the avowed purpose of the federal government to stimulate economic growth coupled with the regulatory mandate to a single purpose authority such as the Environmental Protection Agency or the Occupational Safety and Health Administration. The single purpose authority is ultimately responsible to the president and Congress. Therefore it cannot afford to ignore the economic implications of its decisions anymore than the Atomic Energy Commission's regulatory staff could. Determining which argument has merit will require a comparative analysis of the two types of authority; that falls beyond the scope of this work.

Bearing in mind that the conclusions of this analysis may be limited to improving our understanding of dual purpose authorities, the Atomic Energy Commission's regulation of commercial nuclear power still offers a very rich case for examination. Commercial nuclear plants have a long history of regulation, thus offering a good information base. Since regulation began with the development phase of the technology, there is the opportunity to explore what issues were important and how control was exercised at different stages in its commercial evolution. Nuclear technology also is very complex, and therefore provides a good test of the regulatory process's ability to develop information and resolve technical differences of opinion.

Some would argue that radioactive materials and technologies that use them pose unique threats to the public welfare, indeed to man's ability to survive, and therefore evoke a unique response. Unfortunately, the use of nuclear material does not appear to be such a unique case. Other products and processes are being developed and used, the consequences of which may be similarly catastrophic. These include the effects of fluorocarbons and other man-made chemicals on the ozone layer of the upper atmosphere, attempts to modify weather patterns, and research on recombinant DNA.

The regulatory history of commercial nuclear power, then, is likely to provide insights into the regulation of a broad range of products and should provide us with a clearer understanding of how the regulatory process actually functions, of how technical information is generated and used, of how uncertainty is dealt with, and of how the process deals with change.

Organization

After introducing the reader to a very modest store of technical information necessary to understand the basic structure of a nuclear reactor and the hazards it may pose, the story unfolds more or less chronologically. Chronology is the backbone of this analysis because the regulatory history of the commercial reactor seems naturally to separate into six quite discrete periods, each illuminating its own set of analytical questions. There are the pre-1954 years, when the rudimentary guidelines of reactor safety evolve as both test and military reactors are built. Between 1954, when Congress gives the Atomic Energy Commission explicit responsibility for safety and for commercial reactor development, and 1963, when the private sector begins buying reactors, organizational and procedural evolution are the dominant themes.

In 1963 New Jersey Light and Power Co. bought the first unsubsidized reactor and demand skyrocketed over the next several years, forcing the Commission to face the issue of making trade-offs between safety and economic development. Between 1965 and 1967 some serious questions arose regarding the safety of the reactor, offering a good opportunity to explore the Commission's response to uncertainty. By the late 1960s, public attitudes toward environmental and safety questions seemed to shift markedly, and how the Commission responded to change becomes the dominant question. Then because the popular folklore blames most economic difficulties besetting the commercial reactor on environmentalists' opposition and because, if true, this fact would reflect poorly on the efficiency of the AEC, I devote a short chapter to exploring the effect of regulation on reactor delays and costs. Finally, a number of the above issues are reexamined as a maturing agency attempts self-reform and fights a losing battle to regain public confidence.

Of course history is not so neat that exploring the analytical questions completely does not require some backtracking or getting a bit ahead of the story. And where necessary, this is done at the expense of chronology. To highlight the analytical questions, they are discussed at each chapter close, while the book's conclusion is an overall assessment of the AEC's regulatory role.

On the basis of this introductory roadmap, we might begin.

Notes

1. R.L. Ashley, ed., *Nuclear Power Siting Report,* Proceedings of a National Topical Meeting, American Nuclear Society, L.A. Section, February 16-18, 1965, AEC, CONF-650-201, p. 15.

2. See L.W. Orr and L.O. Mott, *Journal of Economic Entomology* 38 (1975):428-432.

3. *Webster's New Collegiate Dictionary,* 1975 Ed.

4. For a further discussion of these definitions and their operational differences, see William K. Lowrence, *Of Acceptable Risk* (Los Altos: William Kaufman, Inc., 1976).

2 Nuclear Power: Its Scientific Underpinnings

A star is drawing on some vast reservoir of energy. . . . This reservoir can scarcely be other than the sub-atomic energy which, it is known, exists abundantly in all matters; we sometimes dream that man will one day learn how to release it and use it for his service.[1]   —A.S. Eddington, 1920

Details of the physical and engineering principles underlying nuclear reactor technology are extremely complex and not easily absorbed by the layman. But the fundamentals are readily comprehensible, and it is worthwhile for the student of the nuclear regulatory process to become familiar with them.

As the nineteenth century drew to a close, atomic physicists found themselves on the brink of a new understanding of the nature of matter. One after another, pieces to the atomic puzzle were discovered and put into place. Much of the research focused on radioactive materials. In 1895 Roentgen discovered the X-ray. That was followed in 1896 by Becquerel's discovery of alpha, beta, and gamma emissions from uranium, and in 1897 by Thomson's discovery of the electron. The following year, the Curies isolated radium and polonium, and in 1903 Rutherford showed that every radioactive process resulted in the transmutation of an element.[a]

That radioactive materials could spontaneously emit particles capable of causing ionization and affecting a photographic plate meant energy was being liberated. Even before 1900, the small group working with radioactive materials fully appreciated the possibility that the atom might be a source of usable energy. Then, in 1905, Einstein published his theory of the relationship and convertability of mass to energy. His understanding of this equivalence provided the foundation for understanding the phenomenon of radioactivity and atomic energy in general.

Over the next several decades scientists speculated on the possibilities of a much greater source of energy than radiation: that of joining atoms together, or fusion. But the discussion remained theoretical. Then in the early 1930s the development of methods for accelerating charged particles led to discoveries of nuclear reactions releasing almost twice the energy emitted during the disintegration of radioactive atoms. But accelerating sufficient charged particles to cause the reaction required far more energy than was released. The atom was still not a promising source of energy.

---

[a]For a more detailed description of the history and scientific principles underlying nuclear technology, see appendix A.

11

Early in 1939, the outlook changed dramatically. Several groups of scientists on both sides of the Atlantic independently verified the phenomenon of nuclear fission. Uranium nuclei were bombarded with slow or "thermal" neutrons. The nuclei split, releasing several million times the energy carried by the incident neutrons. Moreover, in the process of dividing, each uranium nucleus released two to three neutrons, each itself capable of splitting another uranium nucleus. Thus, the reaction not only produced previously unimagined energy but it also appeared to have the potential of self-propagation. In previous fission experiments, one incident particle could cause the transmutation or splitting of only a single nucleus. Then the reaction was complete. Now it seemed possible that one neutron could lead to the fission of all the uranium in the target.

Since nuclear processes occur very rapidly, there clearly was great explosive potential in a nuclear chain reaction delivering such quantities of energy. American scientists working in the area were immediately aware of the military possibilities and decided, in the interest of national security, to withhold publication of any new discoveries relating to nuclear fission. As the United States was drawn inevitably closer to war, President Roosevelt authorized the Manhattan District Project. In December of 1942, the first fission chain reaction was triggered in a reactor built in the squash courts under Stagg Field at the University of Chicago. Shortly thereafter, the first bomb was successfully tested at Alamogordo, New Mexico. The devastation of Hiroshima and Nagasaki followed in August of 1945.

While the military potential of nuclear power was of paramount concern in the years immediately following the war, work also began on the "domestication" of nuclear power. During the war, researchers in the Manhattan District Project had raised the possibility of developing a nuclear reactor capable of generating electricity, but they had little opportunity to pursue the idea.[2] In 1948, however, the newly formed Atomic Energy Commission created a Division of Reactor Development to further explore the technical and economic feasibility of such a reactor. Even at this early date, scientists had a clear grasp of the fundamental ingredients that were needed to make up this complex technology.

Nuclear Power Technology[3]

In fact, the fundamentals of nuclear power technology are not really as esoteric as one might expect. To get usable power from a nuclear chain reaction, that reaction must produce enough heat to make steam that can drive a turbine. It must also be both sustainable and controllable.

When fossil fuels are the power source, the fuel is burned or oxidized and in the process gives off energy as heat. The heat is then used to convert water to steam and the steam drives turbines. Power units based on nuclear fuel operate in much the same way, though the energy comes from nuclear rather than molecular change and it is produced in much greater quantities per volume of fuel. For example, the fissioning of *one pound* of fissile material produces the same heat as the burning of 1,400 tons of coal.

The Chain Reaction

Some heavy elements, like plutonium or uranium, are *fissionable.* That is, they split into two roughly equal parts upon absorbing a neutron.[4] Most *fissionable* nuclei must be struck by fast neutrons before they will split, but fast neutrons do not generally split nuclei as well as do slow neutrons. Therefore, the best fuel for a chain reaction is material that can be split by slow neutrons. Certain isotopes of the fissionable elements have this property; they are called *fissile isotopes* (U-235, U-233, and Pu-239 are the most common). Some other nonfissile isotopes of these same fissionable elements can capture neutrons, but rather than split, they decay and become new fissile nuclei. For example, Uranium-238 becomes Plutonium-239 or Thorium-232 becomes Uranium-233. These isotopes are called *fertile.* A chain reaction can be fueled by fissile material or some combination of fissile and fertile isotopes.

The basic fission reaction results when a fissile nucleus absorbs a neutron and splits into two lighter atoms called *fission products,* and at the same time emits two to three free fast neutrons and some 200 million electron volts of energy. Ninety percent of that energy occurs in the form of heat. The fission products are usually unstable and decay, producing radioactivity and more heat.

A chain reaction occurs when the quantity and configuration of the fuel is such that at least one free neutron, on average, from each fission reaction goes on to split another nucleus. To sustain a *controlled* chain reaction, no more than one neutron, on average, can go on to split another nucleus. If more than one neutron, on average, goes on to split other nuclei, the rate of the reaction increases rapidly, and it runs out of control.

The Reactor

Six fundamental components make up a nuclear reactor (see figure 2-1). Fuel for the reactor must include an adequate amount of fissile material. The only fissile material occurring naturally in significant quantities is U-235. This is found in natural uranium ore, 99.3 percent of which is nonfissile U-238. The percentage of U-235 can be increased by "enrichment." Other fuels could include U-233 converted from Thorium-232 or Plutonium-239 converted from U-238.

The reactor is usually fueled with both fissile and fertile material, since varying amounts of new fissile material can be made, depending on the reactor

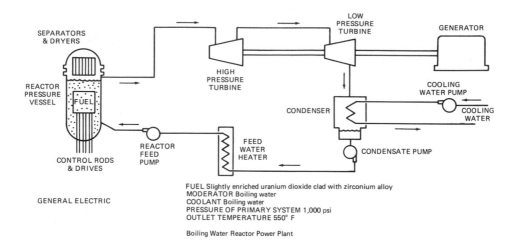

FUEL Slightly enriched uranium dioxide clad with zirconium alloy
MODERATOR Boiling water
COOLANT Boiling water
PRESSURE OF PRIMARY SYSTEM 1,000 psi
OUTLET TEMPERATURE 550° F

Boiling Water Reactor Power Plant

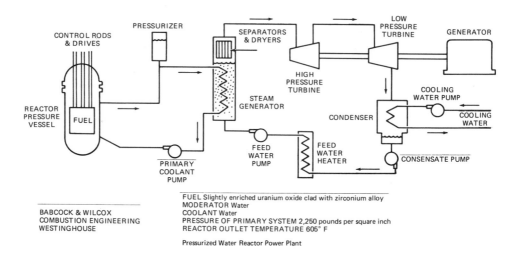

FUEL Slightly enriched uranium oxide clad with zirconium alloy
MODERATOR Water
COOLANT Water
PRESSURE OF PRIMARY SYSTEM 2,250 pounds per square inch
REACTOR OUTLET TEMPERATURE 605° F

Pressurized Water Reactor Power Plant

Source: Wash-1345.

Figure 2-1. Commercial Power Reactor Schematics

design. Although some reactors use fast neutrons, most rely on the slow neutrons and must have a moderating material to slow the newly emitted fast neutron down. Light water, heavy water, and graphite are most commonly used.

Third, a reactor must have a coolant to transfer the heat away from the reactor core and to the steam generator. Water, serving as both moderator and coolant, is most commonly used. But liquid metals, carbon dioxide, and helium gas have also been used.

Fourth, means must be provided for controlling the rate of the reaction. This is done by using rods of neutron-absorbing materials, like boron or hafnium, that can be inserted or withdrawn to control the number of free neutrons.

Fifth, a reflector surrounds the reactor core to reflect back neutrons that might otherwise escape.

And, finally, there is the reactor vessel, which must completely and reliably contain the above components under the stresses of extreme generating temperatures and pressures and of prolonged exposure to radiation.

For use as a source of power, the reactor or nuclear steam supply system simply replaces the fossil fuel burner as a source of heat. The balance of the generating station or the propulsion system remains basically the same.

Fuel Cycle

Of no less consequence to the viability of power reactor technology is the nuclear fuel cycle. New fuel has to be produced and used fuel must ultimately be disposed of. New fuel comes from mined, natural uranium. Uranium ore typically contains only .1 percent uranium oxide in a form that can be mechanically extracted. The residual "tailings," while not useful as reactor fuel, continue to be a source of radioactive radon gas. For most reactors, the concentration of U-235 in the separated uranium oxide or "yellow-cake" must be increased by enrichment. Finally, fuel elements are manufactured from the enriched uranium, possibly in combination with additional fertile material and other fissile material recovered from spent fuel elements.

Well before all the fissile material has been consumed in the nuclear reaction, fission products build up in the fuel element, absorbing neutrons needed for the fission process and threatening the mechanical integrity of the element. The very hot and very radioactive fuel elements must then be removed. Spent fuel elements must be cooled and held at the reactor site for about a year, after which time they can be transported. At this point in the fuel cycle there is the option of reprocessing the spent elements to separate out the substantial quantities of residual uranium and newly converted plutonium for reuse as reactor fuel. Or the spent elements can simply be treated immediately as waste.

Finally, something must be done with the radioactive wastes which include all the fission products and materials that, through neutron-capture or contamination, have become radioactive. Some of these wastes will remain radioactive for many tens of thousands of years.

Hazards of Nuclear Technology

It was immediately apparent to the scientific community that radiation could inflict serious damage on the human body. Within six months of discovering

X-rays, experimenters in Roentgen's laboratory began noticing painful burnlike lesions on their hands. The Curies and their coworkers experienced similar, slow-healing skin problems after the discovery of polonium.

Over the first decade of the twentieth century, diagnostic X-ray techniques came into general use. The mechanics and the full consequences of exposure to radiation were not yet generally understood, and no recommended exposure limits were set. After some years, many early radiologists and technicians began developing skin cancer, leukemia, and forms of anemia. About this time, military radiologists in World War I sharply expanded their use of X-ray to locate embedded shrapnel. Their death rate climbed dramatically. Observation of these facts led scientists for the first time to recommend radiation exposure limits.

After World War I, European radiologists formed national Radium Protection and X-Ray Committees to set exposure limits. The committees met and at first set ceilings of between 3,000 and 4,000 roentgens per year.[b] Technicians continued to die, and in 1927 the United States Committee reduced its recommended limit to 300 roentgens per year.

The following year radiologists organized an International Commission on Radiological Protection to agree on a unified set of exposure standards. The National Committee on Radiation Protection, at first a self-appointed committee and later informally affiliated with the Department of Commerce, represented the United States on the International Commission. While these groups had no official legal standing, standards they set guided exposure levels in most Western countries for the next forty years.

Although scientists still did not understand the mechanisms by which radiation caused damage and disease, they assumed that there was some level of exposure below which the body experienced no ill effects. They further believed a mid-range for exposure existed where the body was damaged but could repair itself. Thus a worker could occasionally be exposed to a demonstrably harmful dose of radiation.

As the years passed, and measurement and experimental techniques improved, effects could be detected at ever lower exposure levels. Recommended exposure limits were consequently reduced by approximately half every five years or so until 1960, when the measurement system was revamped. (See chapter 7 for a full discussion.)

The unique danger of nuclear power technology lies in the fact that it is a potential source of harmful amounts of radiation. Radiation from the reactor or associated fuel cycle operations (mining, milling, enrichment, reprocessing, and waste disposal) enters the environment in four ways. Under ordinary operating conditions, some small amount of radiation is bound to escape any plant using radioactive materials. As the fission process proceeds in a nuclear reactor, the

---

[b]A *roentgen* is that quantity of *x* or gamma radiation that will produce ions carrying one electrostatic unit of electricity of either sign in 1 cc of dry air. A *rem* is the unit dose of ionizing radiation that gives the same biological effect as that of a roentgen of X-rays.

highly radioactive fission products build up. Some—especially those in a gaseous state—inevitably escape the fuel packaging into the coolant. More are also formed in the cooling system by irradiation. In turn, some of these products escape the primary cooling system through small holes in valves, pumps, and seals and move into the outside environment, where they may begin to concentrate—especially in the food chain. Reprocessing plants face similar problems.

Accidents are a second potential source of public exposure. An accident or system failure in a plant using radioactive materials could open the way for a major release of radiation. In a nuclear reactor, an accident or system failure could also lead to a loss of control and a runaway chain reaction. (For descriptions of several kinds of accidents, see figure 2-2.)

Nuclear wastes present a third source of contamination. These wastes are produced at virtually every step in the fuel cycle. Some low-level wastes, for instance, contaminated clothing or tools, are not very dangerous and can be easily buried or otherwise disposed of if their volume is not too great. On the other hand, high-level wastes including spent fuel and reprocessing wastes, present an extreme danger. They are exceedingly toxic and frequently very long lived. For instance, plutonium-239, one of the most lethal substances known to man and an inevitable contaminant of spent fuel, has a half-life of 24,000 years.

A final source of danger lies in the possibility that radioactive material might be diverted from the reactor cycle and intentionally misused either as an explosive or simply as a contaminant.

It was clear to all involved, even in the early years of reactor development, that harnessing the vast energies of the atom was not without its risks. Any emerging nuclear industry would be called upon to design and manufacture components, systems, and plants the reliability of which could be unconditionally guaranteed. Institutions or techniques would have to be developed that could insure the isolation of high-level wastes over many thousands of years, and unprecedented security measures would be required to insure that misuse of radioactive materials was impossible.

Design Features

Nuclear reactor a device in which a fission chain reaction can be initiated, maintained, and controlled. A *light water reactor* (LWR) is one which is cooled and moderated by ordinary water. The function of the moderator is to slow down fast neutrons that are produced in fission, since slow neutrons are much more likely to cause fission in a uranium-235 nucleus. The two basic types of LWRs in commercial use in the United States today are the pressurized water reactors (PWRs) and the boiling water reactors (BWRs). Both types are fueled with uranium that has been enriched to contain 2 to 4 percent uranium-235. The fuel is sealed in tubular metal (zircaloy) cladding, the fuel rods.

BWR *boiling water reactor*; a LWR in which the water *is* allowed to boil in the pressure vessel. The steam generated in the pressure vessel passes directly to the turbine generators, the 'direct cycle' system.

As of June 30, 1975, there were 23 commercial BWRs in operation or licensed by the NRC in the United States, and 53 BWRs under construction, on order, or planned.

PWR *pressurized water reactor*; a LWR in which the water is *not* allowed to boil any place in the pressure vessel. It is kept under high pressure and high temperature to keep it liquid. The heat is transferred to a separate stream of water which boils and produces steam for the turbine. All PWRs employ this dual system for transferring energy from the reactor fuel to the turbine and are called 'indirect cycle' systems.

As of June 30, 1975, there were 30 commercial PWRs in operation or licensed by the NRC in the United States, and 121 PWRs under construction, on order, or planned.

Pressure vessel the largest and heaviest piece of equipment in a nuclear power plant which houses the fuel *core,* mechanisms for driving control rods into and out of the core, and pipes for circulating the cooling water.

Downcomer cylindrical passage between the reactor core and pressure vessel through which the cooling water circulates downward from inlet nozzles before passing upward through the core.

Safety Mechanisms

Scram emergency shutdown system for rapidly terminating the reactor chain reaction. When a planned or unplanned event or accident occurs wherein the reactor has to be shutdown, the performance of a number of systems is involved; their failure could lead to serious problems, including loss of cooling.

ECCS *emergency core cooling system(s)*; the fission reaction that produces most of the heat in a nuclear reactor can be turned off at any time by inserting control rods, containing strong neutron-absorbing materials, into the reactor core. However, the heat produced by radioactive decay cannot be turned off. Thus, a system must be provided to cool the core and to prevent the fuel rods from melting, with potentially catastrophic results, if normal cooling is interrupted. This is the emergency core cooling system, which to date has not been given a full-scale test on a working reactor.

A major forthcoming program at the Idaho National Engineering Laboratory, the loss of fluid test (LOFT), will examine what happens during loss of coolant accidents (LOCA), and how the ECCS behaves in such circumstances at the system test level. The LOFT test reactor is a small PWR designed to carry out LOCA experiments on a size and scale (55 MWt) between that of a laboratory experiment and that typical of a commercial PWR (3,300 MWt).

Source: Primack, Joel, "Nuclear Reactor Safety," *Bulletin of the Atomic Scientists*, September 1975, p. 17. Reprinted by permission of the *Bulletin of Atomic Scientists*, a magazine of science and public affairs. Copyright © by the Educational Foundation for Nuclear Science, Chicago, Illinois.

Figure 2-2. Reactor Safety Glossary

Accidents

Transients events, planned or unplanned, which sometimes lead to a signal to terminate the chain reaction (shut down the reactor). The reactor control system can automatically adjust for some transients without service interruption.

LOCA *loss of coolant accident.* An accident of this sort may range in severity from a catastrophic rupture of the reactor pressure vessel (which is almost certain to lead to a core melt with large potential radiological consequences to the public) to rather minor leaks in pipes (which could be handled by the engineered safety features of the reactor).

Blowdown first phase of a LOCA. The water in the reactor cooling systems is at a very high pressure and if a rupture occurs in the pipes, pumps, valves or vessels that contain it, the rapid pressure release would result in some of the water flashing to steam. The resulting pressure will force most of this water-steam mixture out the broken pipe.

Meltdown if a failure in the reactor's cooling system occurs that allows the fuel to heat up to its melting point (about 5,000°F), a core meltdown would occur.

If the core melts, existing reactor safety systems would be unable to cool it and a core *melt-through* would follow. The molten core would melt through the pressure vessel, containment building, and sink into the earth below (the 'China syndrome').

Containment failure a major release of gaseous and volatile fission products to the atmosphere as a result of the breach of the containment building—the large, dome-like structure surrounding the reactor—following a meltdown.

Reference accident a core meltdown for a 1,000 MWe pressurized water reactor with a subsequent containment failure. This accident, used to set a scale for reactor accident consequences, is the accident for which the most complete information is given in WASH-1400 Draft, and is the accident which the APS study used to compare their analysis with the AEC's study.

Health effects a term used in these articles to denote the consequences to human health of a meltdown followed by a containment failure.

Early, immediate or short-term 'health effects' are fatalities occurring in the first one to two months as a result of acute radiation.

Delayed or long-term 'health effects' are fatalities (cancer deaths) occurring over a period of several decades. Other 'health effects' include genetic defects and thyroid nodules.

Notes

1. A.S. Eddington, Presidential address before the Physical Science Section of the British Association, 1920, as found in Samuel Glasstone, *Sourcebook on Atomic Energy* (New York: Van Nostrand, 1950).

2. Hogerton, "The Arrival of Nuclear Power," *Scientific American* 218:2 (February 1968), 21.

3. This discussion draws heavily on the excellent description of reactor technology found in *Nuclear Power Issues & Choices,* pp. 388-394.

4. Neutrons travel at varying speeds. They are classed as thermal or slow neutrons (<1 ev.), intermediate neutrons (>1 ev. but <1,000,000 ev.), and fast neutrons (>100,000 ev.).

3 The Formative Years

The Commission has been confronted . . . with the dilemma of avoiding a rigid pattern of licensing and regulation which would slow down the development of civilian uses of atomic energy or which would unnecessarily interfere with management practices.

At the same time, the Commission [has] to assure without equivocation, that the public health and safety is protected.[1]

—Joint Committee on Atomic Energy, 1961

The destruction of two major Japanese cities introduced atomic power to the world. Heretofore, this unprecedented source of energy had been only a subject of speculation and some experimentation for an elite handful of the world's physicists. But with its public demonstration, there was no turning back. The question was not whether, but *how,* nuclear energy should be used.

The Atomic Energy Act of 1946

After considerable disagreement over whether control of nuclear energy should rest with a civilian or a military agency, Congress passed the Atomic Energy Act of 1946 (the McMahon Act). The act created a five-man, civilian Atomic Energy Commission, appointed by the president with the consent of the Senate, and declared it

> . . . to be the policy of the people of the United States that, subject at all times to the paramount objective of assuring the common defense and security, the development and utilization of atomic energy shall, so far as practicable, be directed toward improving the public welfare, increasing the standard of living, strengthening free competition in private enterprise, and promoting world peace.[2]

In sum, Congress intended that the United States should continue to guard its weapons monopoly and then, where possible, should advance nuclear technology in general.

To these ends, the Commission was given a virtual monopoly over nuclear technology. All nuclear related programs were transferred from the military to the AEC. The Commission acquired outright ownership of all fissionable materials and of all facilities that used or produced such materials. It also gained control over and placed stringent restrictions on the use of any related technical information.

To give Congress itself more effective oversight control, the act created a Joint Committee on Atomic Energy uniquely empowered to consider and to act on all AEC and nuclear-related legislation. Usually joint committees are authorized only to study problems, while the separate House and Senate committees reserve the power to consider and report on bills. The creation of the Joint Committee on Atomic Energy streamlined the legislative process for AEC legislation and gave the Joint Committee unprecedented power.[3] Although centralized governmental control of facilities, material, and information could not be expected to foster the robust development of a nascent technology, it satisfied perceived national security requirements of the day.

The issue of public health and safety was conspicuous in its absence from the bill. While many of the hazards of radiation were clearly apparent in 1946, no one viewed the potential threat of an evolving nuclear technology as intolerable or something good engineering could not acceptably control. Moreover, since the AEC enjoyed monopoly control over nuclear information and materials, Congress saw little need to regulate private industrial uses.

The one regulatory procedure provided for in the act arose out of congressional fears that fissionable material produced in government installations might be diverted.[4] To guard against this contingency, the act required that the Commission license all fissionable material and its transfer. With specific legal control, the Commission could then prosecute and have the courts impose stiff penalties for violations. As part of the same licensing program, Congress directed the Commission (1) to control the distribution of fissionable material as required to protect the public health and safety, and (2) to license equipment using fissionable material after Congress itself had had ninety days to review the economic and safety implications of introducing the equipment.

Safety in the Civilian Reactor Development Program

In fact, the AEC exercised very limited formal licensing and regulatory control during its early years. It had two small staff offices, one in Washington and one at Oak Ridge National Laboratory, to administer the licensing program. And between 1947, when the McMahon Act took effect, and 1954, when it was superseded, the Commission adopted only two minor safety regulations.[5]

However, its operating responsibilities required the Commission to make decisions that, in effect, informally defined a safety program. First, the commissioners had to settle on some criteria to govern the siting of the new reactors they were charged with developing. Since the reactors were experimental and the engineering, by and large, was untested, they continued the policy of isolated siting used to locate plutonium production and research facilities during the war.

At this early stage in the development of nuclear technology, it was neither sensible nor practical to separate safety research from development research. Research into the physical constants governing the nuclear reaction, the effects of extreme temperature, pressure, and radiation stresses on reactor materials, and the usefulness of various concepts and fuel configurations contributed equally to the development of a successful reactor and of a safe reactor. Thus, in 1947, the Commission assigned all power reactor research to its newly formed Reactor Development Branch.

Reactor technology was very complex and understood by only a few during these early years. In fact, most of the commissioners and the AEC's general manager were lawyers and administrators and not themselves experts in the field. Realizing the Commission would need access to good scientific advice, the 1946 legislation had created a General Advisory Committee upon which the Commission could and did rely heavily for technical advice.

The Advisory Committee immediately foresaw that major safety considerations would inevitably be a part of any serious nuclear development program and recommended that the Commission appoint a second committee to advise it on safety questions. In 1947 the Reactor Safeguards Committee, chaired by Edward Teller, was appointed and given responsibility for insuring the safety of the reactor development program. Three years later, the Commission also appointed a special Industrial Committee on Reactor Location Problems to evaluate potential reactor sites. The two committees were then merged into one Advisory Committee on Reactor Safeguards (ACRS) in 1953.

The Reactor Safeguards Committee gave full support to the Commission's general policy of remote siting. As a further precaution against public exposure to accidental or even normal operating releases of radiation, the Commission approved a committee recommendation that reactors, then designed without any containment superstructure, be further isolated by a restricted zone around the facility. That zone, scaled according to an ACRS formula based on reactor power output, was to be permanently closed to public use.[6] Consistent with these beginnings, the ACRS continued to rely primarily on distance to give the margin of safety it thought necessary.

In 1949 the Reactor Safeguards Committee came up with a second line of defense. General Electric applied to the Commission for permission to build a small power breeder reactor on its Knolls Laboratory site, twenty miles north of Schnectady. As a matter of course, the ACRS reviewed the proposal and recommended that approval be granted if General Electric enclosed the entire reactor facility in a heavy structure "capable of containing any radioactivity that might be produced in a reactor accident."[7] Thus the spherical containment shell—now the trademark of a nuclear reactor—first came to be.

Safety in the Naval Reactor Program

While the AEC was charting a rather deliberate course of reactor research and development, then-Captain Hyman Rickover applied himself in a characteristi-cally single-minded fashion to the job of persuading his superiors that develop-ment of a nuclear-powered submarine should be one of the navy's top priority goals. In 1948, after a year of lobbying, Rickover won assignment as head of the Nuclear Power Branch of the navy's Bureau of Ships. The following year, after the Commission formed its own Division of Reactor Development, Rickover's assignment was expanded to include the post of Chief of the Naval Reactor Branch of the AEC. The two branches, comprising the same group of people, were set up to allow the AEC and the navy to participate jointly in the development of a nuclear submarine (see figure 3-1).

Wearing both hats, Rickover was then able to command the resources and exercise the authority to pursue propulsion reactor development with the commitment he felt it warranted.[8] He wanted a full-scale, operating nuclear-powered submarine in the shortest possible time as proof that the technology was serviceable and in hand. Therefore, after minimal research on alternatives, he settled on Westinghouse's pressurized water reactor design in 1950. It may or may not have been the best long-run candidate, but it was the most advanced at that time. By June 1953, Rickover's land-based prototype, the Mark I, had successfully completed its first 100-hour trial run at the reactor test facility in Idaho.

Rickover was very aware that any serious accident would probably kill support for further reactor development work and therefore paid keen attention to matters of safety. As a member of the Commission staff, he worked closely with the Advisory Committee on Reactor Safeguards on the Mark I prototype and arranged for the committee's step-by-step approval as the unit neared completion.[9] But margins of safety for a nuclear-powered submarine could not be guaranteed in the same way they were with land-based reactors. Since the sixty-man submarine crew had no avenue of escape while the ship was at sea and major ports were generally large population centers, "remote siting" could not be relied upon to acceptably limit the consequences of an accident. Nor could containment be reasonably engineered for a submarine.

Instead of protecting against the consequences of an accident, Rickover had to guarantee that the accident would never happen. To do this, the engineering and reactor design had to be error-free, the components and assembly flawless, and the operating personnel totally competent and reliable. Protection lay in quality, not safeguards. Rickover committed himself to these goals. He and his group carefully reviewed and synthesized nuclear-fission research. He personally oversaw the design and manufacture of the reactors by Westinghouse, setting a

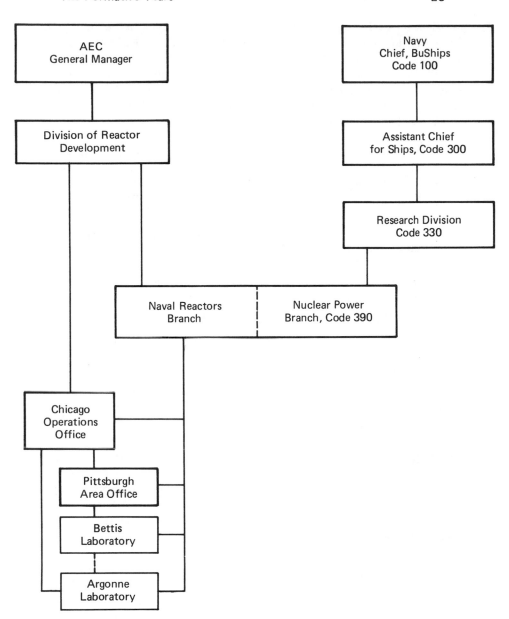

Figure 3-1. The Navy Nuclear Propulsion Projection in January 1949

precedent for quality assurance in the naval reactor program that far outstripped that found in either AEC or later commercial projects.[10] Rickover kept and personally reviewed detailed operating records to correct incipient problems.[11] And finally he personally selected his operating personnel and then put them through a grueling training in reactor design and operation. He viewed their reliability and competence as the ultimate safety barrier.[12] Describing his safety program to the Joint Committee, Rickover testified that

> ...the Naval Reactors Branch has had an in-house headquarters organization responsible for the design and development of reactors, the construction of the nuclear power plants that contain the reactors, the training of officers and men, and for the constant observance of all the major and minor technical details that occur. We are involved in these details and we follow them thoroughly on a day-by-day basis. This is the way we make sure that our reactors are designed and operating properly.[13]

And so it was that in these very early years of reactor development two quite divergent approaches to reactor safety evolved. On the one hand, the AEC, supported by the ACRS, chose to rely ultimately on techniques that limited the consequences of an accident. On the other, Rickover chose to rely on meticulous quality control—of design, of manufacture, and of personnel—to limit the probability of an accident. The naval and the civilian programs continued to be carried on independently, so there was no cause for immediate conflict. But in later years, as men schooled in the Rickover approach moved into policy-level jobs with AEC, a bitter contest between exponents of the two positions arose.

The Atomic Energy Act of 1954

The civilian sector was somewhat slower than the military to perceive a future for nuclear power. But during the early 1950s, a variety of manufacturers and electrical utilities put together study teams to explore the potential of nuclear power generation. The teams concluded that, assuming resale of by-product plutonium to the military, nuclear plants could generate electricity at a cost of roughly 7 mils per kilowatt-hour. At that time, conventional power production costs ranged from 5 to 10 mils per kilowatt-hour and averaged 7 mils.[14]

Responding to industrial interest, the Division of Reactor Development shifted its development priority to the central power station reactor as the naval reactor program moved closer to a viable product.[15] Then, in the spring of 1953, Commission Chairman Gordon Dean urged the Joint Committee on Atomic Energy to make the development of a nuclear generating capability a top government priority. To facilitate private participation in such a development program, Dean also recommended that the McMahon Act be amended to permit

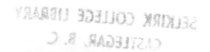

(1) private ownership of nuclear reactors; (2) private ownership or leasing of fissile material; (3) more liberal patenting; (4) general industrial access to information needed to make sound design and engineering decisions; and (5) imposition of appropriate safety and siting regulations.[16]

At the same time the private sector began showing some interest in developing nuclear power, events in the postwar world were fast rendering the underlying objectives of the McMahon Act obsolete. In August of 1953, the Soviet Union tested its first thermonuclear device, demonstrating a practical command of nuclear technology. With the loss of our nuclear weapons monopoly, the early emphasis on security became less pressing. At the same time, the cold war conflict moved from confrontation in Europe into a new phase of competition for allies among uncommitted countries, and the United States was eager for opportunities to demonstrate superior technological prowess.

Responding to these changes, President Eisenhower presented his new "Atoms for Peace" initiative to the United Nations in December of 1953. He called for international cooperation in developing peaceful applications of nuclear energy, particularly by nuclear power. The international development effort was to be marked by an effective system of international control over the use of fissionable materials.[17] In his budget message a month later, the president recommended that the Atomic Energy Act of 1946 be amended to permit sharing information and fissionable materials with friendly nations and to encourage U.S. industry to develop nuclear power.[18]

For its part, the Republican-dominated Joint Committee on Atomic Energy had long supported an expanded commercial reactor development program with private participation. In fact, the committee had begun drafting amendments to open the industry to private participation as early as 1951, but at that time it got no support from the Commission.[19] By 1954, the Commission, the president, and the industrial community, all for somewhat different reasons, joined together in support of overhauling the 1946 legislation, and the Joint Committee welcomed the opportunity. Hearings were begun in the spring of 1954 and the new Atomic Energy Act of 1954 was signed by the president that same year.

The overriding policy objective to be met in amending the McMahon Act was to facilitate the commercial development and exploitation of nuclear power by private industry. The key provisions of the new legislation permitted (1) private ownership of nuclear facilities, (2) private use of fissionable material (though the AEC still retained title to it), (3) liberalized patenting rights, and (4) industrial access to needed technical information. The 1954 act also directed the AEC to develop those programs necessary to introduce nuclear energy into the private sector.

The regulatory objectives most debated during passage of the new legislation and given primary consideration in the act itself centered on making the nuclear

industry economically independent and internally competitive. No questions
regarding the possible safety hazards of nuclear technology were ever explored in
the hearings and there were no discussions of what might constitute an
"acceptable" level of risk.[20]

Although the hearing record indicates that nuclear safety was not an issue in
1954, the Joint Committee clearly recognized that nuclear technology posed a
unique threat to the public welfare, and that any proposal to move it into the
public domain must include some regulatory provisions.[21] In similar situations
where the regulatory choices were politically thorny or demanded some measure
of expertise, Congress had traditionally passed responsibility to the appropriate
commission or executive department. Nuclear power was to be no exception.

The regulatory provisions of the 1954 legislation defined a licensing system
modeled on the 1935 Federal Communications Act.[22] The AEC received a clear
mandate to

> prescribe such regulations or orders as it may deem necessary . . . to
> guard against loss or diversion of any special nuclear material . . . to
> govern any activity authorized pursuant to this Act including standards,
> design, location, and operation of facilities used in the conduct of such
> activities in order to protect health and minimize danger to life and
> property.[23]

The act also required the Commission to license all facilities and operators
producing or using radioactive materials. In the case of facilities built to use
nuclear materials (that is, central power stations), the provisions were somewhat
more detailed. First, construction permits were required, and

> all applications for licenses to construct or modify production or
> utilization facilities shall, *if the application is otherwise acceptable,* be
> initially granted a construction permit.[24]

Then,

> upon finding that the facility authorized has been constructed and will
> operate in conformity with the application as amended and in con-
> formity with the provisions of this Act *or of the rules and regulations
> of the Commission,* and in the absence of any good cause being shown
> to the Commission why the granting of a license would not be in
> accordance with the provisions of this Act, the Commission shall
> thereupon issue a license to the applicant.[25]

Administrative procedures set down in the act required that applications be in
writing and that those for operating licenses give such information as

> the Commission may by rule or regulation deem necessary in order to
> enable it to find that the utilization or production of special nuclear

materials will be in accord with the common defense and *will provide adequate protection to the health and safety of the public.*[26]

A further provision required that in any proceeding for issuing licenses or regulations "the Commission shall grant a hearing upon the request of any person whose interest may be affected by the proceeding,"[27] and that "any final order . . . shall be subject to judicial review."[28]

These provisions spelled out the barest essentials of a regulatory process. The AEC was responsible for adequately protecting the public health, safety, life, and property. The means resided in the AEC's licensing power and its authority to make rules, compliance with which were prerequisite to licensing. Provisions for public hearings and judicial review allowed some opportunity for redress in the event of abuse of power or dereliction of duty. But they gave no guidance on what the criteria or standards for licensing should be. Nor did they in any other way define the point where the risk imposed by the production or use of radioactive material became an unacceptable threat to the public welfare.

Since Congress chose not to define "adequate protection" or "undue risk," that job fell to the AEC. The Commission had the choice of making the definition explicit, perhaps through some process of risk analysis, or implicit, through licensing standards. Since both the technology and the analytical methodologies were in their infancy, determining where risk became "undue" was left to be hammered out over time through the regulatory process.

Notes

1. Joint Committee on Atomic Energy, *Radiation Safety and Regulation,* June 1961.

2. P.L. 585, 79th Congress (60 Stat. 755).

3. For a full discussion see Green and Rosenthal, *Government of the Atom* (New York: Atherton Press, 1963).

4. Berman and Hydeman, *The Atomic Energy Commission and Regulating Nuclear Facilities* (Ann Arbor: University of Michigan Law School, 1961), p. 65.

5. Joint Committee on Atomic Energy, *Improving the AEC Regulatory Process,* Volume II, March 1961, p. 438.

6. Frank G. Dawson, *Nuclear Power: Development and Management of a Technology* (Seattle: University of Washington Press, 1976), p. 176.

7. Hewlett and Duncan, *Nuclear Navy* (Chicago: University of Chicago Press, 1974), p. 176.

8. For a full discussion of this unusual dual role see Hewlett and Duncan, *Nuclear Navy,* pp. 86-92.

9. Ibid.

10. Arthur W. Murphy, ed., *The Nuclear Power Controversy* (Englewood Cliffs, N.J.: Prentice-Hall, Inc., 1976).

11. Ibid., p. 365.

12. Joint Committee on Atomic Energy, *Hearings on Radiation Safety and Regulation,* June 1961, p. 361.

13. Ibid., p. 359.

14. J. Hogerton, "The Arrival of Nuclear Power," p. 28.

15. Ibid., p. 23.

16. Wendy Allen, *Nuclear Reactors for Generating Electricity: U.S. Development from 1946-1963,* The Rand Corporation, R-2116-NSF, June 1977, p. 36.

17. Hogerton, "The Arrival of Nuclear Power," p. 23.

18. Green, *Government of the Atom,* p. 124.

19. Ibid., p. 118.

20. Berman and Hydeman, *The Atomic Energy Commission . . . ,* p. 68.

21. Ibid.

22. Private communication, Hewlett, December 1976.

23. P.L. 83-703 (68 Stat. 1919), 1954, Chap. 1.

24. P.L. 83-703 (68 Stat. 1919), 1954, Chap. 14, Sec. 161. Emphasis added.

25. Ibid., Sec. 185. Emphasis added.

26. Ibid. Emphasis added.

27. Ibid., Sec. 182.

28. Ibid., Sec 189 (a) and (b).

4 The Ground Rules

I believe that any professional group should have the privilege of privacy in arriving at technical decisions in their particular fields.[1] —Chauncey Starr, 1965

There are four conceptually identifiable ingredients in a functioning regulatory program: (1) the regulatory objective(s), (2) standards or criteria by which the technology is made to conform to those objective(s), (3) procedures for setting and applying standards fairly, and (4) enforcement procedures and sanctions. The Atomic Energy Act of 1954 specified some of these ingredients, but left much to the judgment of the Commission.

Congress defined the regulatory objective as providing "adequate protection to the health and safety of the public" or more ambiguously as "[protecting] health and [minimizing] danger to life and property." If the objective was to provide adequate protection, then "adequate" needed definition.

Since no definition was attempted in the legislation, it fell to the AEC to establish criteria or standards that would define "adequate." These could be formal standards, publicly adopted by the Commission as a prerequisite to the operation of a facility. They could be informal criteria, customarily used or recommended by the regulator or by some other participant in the regulatory process.[a] Or they could be a function of the compliance process where consistent lack of enforcement is a time-honored technique for amending regulations or relaxing standards.

Just as standards could take many forms, they could be based on information from many different quarters. Before an AEC standard could be adopted, two separate assessments needed to be made. First, if one applied that standard, what risk would be incurred? That is, under the specified design or standards, what was the likelihood of radiation escaping and if radiation did escape, what effect would it have on health and property? Second, was that effect acceptable? In a reasonably open, democratic system, the risk assessment, even if it is subjective and intuitive, must have some external basis or support. It could be based on expert opinion or testimony, on reflections of popular perceptions, or on specifically tailored research programs.

To avoid burdening a new industry with any unnecessary costs or constraints, regulatory requirements needed to be set at just that level where the

---

[a]An example of this latter situation arises in the case of the intervenors who were, in some cases, consistently able to impose specifications on nuclear plant design that were more exacting than AEC requirements. See text following.

risks became "acceptable"–presumably to the group at risk. But how was that
level to be determined? In practice, technical experts were often relied upon to
make judgments of "acceptable risk level" at the same time they made risk
assessments, though they had no special qualifications for doing so. Alternatively, the AEC set standards based on its own perceptions of the public will and
then relied upon various feedback mechanisms to validate its judgment. Feedback came from a variety of sources including lobbyists, intervenors in the
regulatory process, and congressional attempts to override Commission decisions.

By contrast, Congress was more forthcoming in delineating the regulatory
procedures by which standards should be applied to applicants for licenses. The
standards or criteria by which a facility was judged safe had to be prescribed in
formal regulations or Commission orders subject to the normal practices of
administrative procedure.[b] The Commission then judged whether or not the
applicant was in conformity with the standards, and if there was a disagreement,
it was resolved on the basis of a formal public hearing record. If disagreement
persisted, it could ultimately be resolved in the courts.

While the legislation clearly required that certain elements be included in
the regulatory process, much of the substance of the process was still left to the
discretion of the Commission. No procedures were specified for arriving at rules
or standards, nor was the process described by which the Commission initially
decided a permit or license should be granted.

Finally Congress charged the AEC's Inspection Division with verifying that
contractors, licensees, and employees of the AEC comply with the provisions of
the act and with the rules adopted by the Commission.[2] To put teeth into the
compliance provisions, Congress also provided for sanctions of up to a $10,000
fine or five years in prison for willfully violating the licensing and use provisions
of the act.[3] But this authority represented an upper bound of enforcement. As
with all policing authorities, there could be a substantial gap between the
allowable and the actual sanctions imposed, and how large a gap would pretty
much be a matter of Commission discretion.

Although the Commission enjoyed substantial freedom in deciding how to
flesh out the regulatory program required by Congress, certain technical and
political realities of the late 1950s constrained its choices to some degree. During
this period the AEC came under increasingly intense pressure to develop a
commercially viable power reactor. The president, the Joint Committee, and the
new Commission chairman, Admiral Lewis Strauss, had come to support
governmental intervention to make that objective a reality. It seemed clear this
group would oppose any vaguely grounded considerations that might seriously
slow progress toward its objectives.

---

[b]These practices required that a notice of intent and the full content of a proposed
Commission Order on rule be published in the Federal Register at least thirty days prior to
adoption. During this period any opponents of the ruling could present their case to the
Commission.

While they were unified in their support of commercial nuclear power, the key political participants diverged sharply on details of how to bring it into being. The battle was an old but hotly contested one between the Democratic advocates of government ownership of generating and distribution facilities and the Republican supporters of private control. The Joint Committee was divided between the two almost equally, while the president and his appointees to the Commission unanimously defended the role of private enterprise. Conflict over this issue would not necessarily have had to affect the regulatory process. But, in fact, the animosities were such that the participants struck out at each other wherever possible. Inevitably they had occasion to lock horns over regulatory questions.[4] (See discussion of the Fermi reactor following.)

Although it became the focus of a major governmental development effort, power reactor technology remained, for the most part, experimental during these years. No single type of reactor emerged as holding the greatest commercial promise. And under the newly launched, five-year Power Demonstration Reactor Program, the AEC subsidized the development of at least six different design alternatives.[c] And although power reactors were certainly not economically competitive with fossil plants during the late 1950s, Commonwealth Edison, Consolidated Edison, and Pacific Gas and Electric each ordered a reactor without any direct government subsidy. These three utilities believed nuclear technology showed great promise, and they were willing to make early investments to develop staff expertise in nuclear construction and management.

All the commercial projects had several characteristics in common. They were small, ranging in capacity from 10 MWe to 255 MWe.[d] They were few; utilities submitted only fourteen nuclear plant construction permit applications between 1955 and 1962. Each one had a unique reactor design and a unique site. And the AEC's Division of Reactor Development worked closely with each manufacturer and with each utility.

Organization of the AEC

To accommodate the changes in regulatory and development functions required by the Atomic Energy Act of 1954, the Commission created a new Division of Civilian Applications, initially headed by Harold Price, a lawyer. Most development responsibilities remained with the Division of Reactor Development. However, the new division was assigned all regulatory and a modest complement of promotional responsibilities, a combination not then perceived as holding any inherent conflict of interest. These responsibilities included (1) administering a

---

[c]These included the boiling water, the fast breeder, the heavy water, the organic moderated, the pressurized water, and the sodium cooled graphite moderated designs.

[d]MWe mean megawatts of electrical energy the plant is able to generate. A given amount of electrical energy is generated by roughly three times that amount of heat energy. Therefore a 100 MWe plant will be approximately the same as a 300 MWth plant.

program to provide industrial access to classified information, (2) pricing government-provided nuclear materials and services, (3) initiatives to encourage private use of nuclear technology, and (4) the development of a comprehensive regulatory program. While the Division of Civilian Application was responsible for regulatory control, the Division of Reactor Development offered substantial advice on questions of reactor design and on technical specifications under consideration in the licensing process.[5] As so aptly noted in a Joint Committee study,

> ...the Commission has established an organization and has allocated responsibility among its divisions in a way that is tailored to the demands of the present state of the atomic power industry, in which the regulatory work is comparatively light, and is substantially inter-related with the developmental and promotional programs.[6]

In April of 1955, the Commission had set up a Hazards Evaluation group as part of the general manager's staff.[e] This group, composed initially of only two people, was charged with evaluating all applications for construction permits and operating licenses.[7] Once the Division of Civilian Applications was created, the Hazards Evaluation group came under its jurisdiction, retaining responsibility for licensing activities. The group, working closely with the applicant, assessed the safety characteristics of the reactor components and overall design. It then presented its conclusions in a staff report to the division director, to the Advisory Committee on Reactor Safeguards, and to the Hearing Examiner if the decision went to a hearing. While the two-man Hazards Evaluation staff was ultimately responsible for assessing the safety of an applicant's design, in the beginning it relied heavily on the greater expertise and experience of the fifteen-man Advisory Committee on Reactor Safeguards.[8]

Because the law required commercial reactors to be licensed prior to construction and again prior to operation, the Commission moved with some dispatch to organize a licensing capability. By contrast, the need for an explicit set of guides or standards seemed less pressing, and no office or group was set up to develop any. No specifically safety-oriented research program was organized, either. Instead, the Hazards Evaluation staff and the Commission relied on the ongoing research programs of the Divisions of Reactor Development and Biology and Medicine.[9]

The organization of the regulatory function within the AEC suggested what proved to be the case during the early years of the program. The agency focused on the expeditious processing of license and permit applications. It did not, at that time, choose to commit resources to a standards or criteria development program. Nor did it perceive any need for an independent research program responsive to the needs of the regulatory program.

---

[e]The general manager was the AEC's chief executive officer. He was appointed by the Commission.

The Commission's priorities were corroborated in 1955 by two independent evaluations of the AEC's organization. Both the management consulting firm, McKinsey & Co., and the Joint Committee submitted reports stressing the need for "facilitating the task of potential licensees." Neither offered substantive evaluations of safety and health objectives or of the AEC's responsibility in that domain.[10]

The Regulatory Process

The Commission moved quickly to adopt the regulations necessary for a smoothly functioning licensing process. To assure a broad base of support for its proposals, the Commission worked closely with an advisory committee of state officials and with various private groups. As the proposed regulations reached the final form, the Commission also provided all interested parties with an opportunity for written public comment. By mid-1955, the Commission formally proposed the regulations that would govern the licensing of special nuclear materials and use facilities and access to classified information. By 1956 these regulations were in effect and the Commission was working on a draft of basic radiation protection standards.

Typically, an applicant (the utility in the case of the power reactor) would have his first agency contact with the Division of Reactor Development to discuss the availability of fuel and basic plant design. If the applicant sought assistance under the Power Reactor Demonstration Program (managed by the Division of Reactor Development), he might then receive a formal and public AEC commitment to the project—although no license had been applied for.

Generally the applicant's first contacts with the Hazards Evaluation Branch were to get informal approval for his proposed site. But even by this stage he had made a significant investment in gathering information on the proposed site and the reactor design. Once the applicant had firm commitments for a site, a reactor manufacturer, a plant design, and financing, he filed a formal application with the AEC for a construction permit. The application included a preliminary safety analysis report (PSAR) that described the site, the reactor, operating procedures, staff training requirements, and fuel transport and storage procedures. The staff also required the applicant to include an accident analysis demonstrating that any "credible accident" would not expose neighboring populations to more than the exposure levels established by the AEC in Chapter 10, Part 20 of the *Federal Register* (10CFR20).

Because application for a construction permit had to be made early, the applicant's design analysis tended to be incomplete, often containing elements still in the research stage. The staff of the Hazards Evaluation Branch therefore could only make a subjective assessment of the plant and site combination and the likelihood that the unresolved design problems could and would be favorably resolved. To recommend issuing a construction permit, the staff required reasonable assurance that further study would favorably resolve design problems or that alternative solutions already existed.

Once completed, the Hazards Evaluation staff report was forwarded with the original application to the Advisory Committee on Reactor Safeguards for a second review. Often the Advisory Committee apprised itself of the basic information on a project even before an application was formally submitted, but committee members had no tradition of informal contact with the applicant prior to review. The committee had only a small staff to assist with its growing workload and therefore came to rely heavily on its own subcommittees and outside consultants for substantive reviews.[11] Members would meet monthly for two to three days and were able to finish a review in two to three months. They arrived at their final evaluation in private. And since the committee was not statutory but existed at the pleasure of the Commission, its report to the Commission was private and subject to executive privilege.

Once the Hazards Evaluation staff report and the Advisory Committee reports were submitted, the Commission decided to issue or deny a permit— accepting the advice of its staff and the Advisory Committee or not, as it chose. There was no public record of safety problems that might have been raised or reasons for the Commission's decision.

The Commission then gave formal notice of action taken by publishing the construction permit in the *Federal Register*. Any interested party then had thirty days during which to request a hearing before a hearing examiner to establish a formal record. Because the notice was of *action taken* and not of a *proposed decision,* the applicant could proceed with construction during the thirty-day period. In fact there was no explicit Commission ruling on how far construction could proceed before a construction permit had to be issued.

The construction permit entitled the utility to build a plant on a particular site, by a specified date, and with the design characteristics specified. Once the construction permit was issued, the utility began the jobs of more advanced research on components, detailed design, and construction. Typically, all were in progress simultaneously. As the plant neared completion and the details of the design and operation finally became fixed, the utility could apply for an operating license.

There was little doubt at this point that the plant would ultimately be granted permission to operate, but the operating license evaluation was necessarily more detailed and time-consuming than that for the construction permit. The more difficult, unresolved problems had been inevitably deferred at the construction permit stage, and the sheer quantity of detail in the plant design and the operating procedures required time to evaluate. Once the staff approved the plant (and if no one requested a hearing), the examiner issued an operating license to the utility specifying permissible power levels and operating characteristics for the reactor. Licenses were generally valid for fifty years. If the utility wanted to begin preliminary reactor fueling and testing when some details remained unapproved, the examiner could issue a temporary operating license that the utility later had to convert to a permanent license.

The rationale behind dual licensing requirements was straightforward. The Joint Committee believed that the investments a utility must make in land acquisition and design were so great that, without some formal approval early in the application process, no utility would risk the capital required to move to the operating license stage. So requirements for both a construction and operating license were written into the 1954 act.

This approach clearly had its own problems, particularly in the early years of reactor development. Nuclear technology was evolving so rapidly that construction permit applications could include only the most general description of the reactors. The plant was virtually complete before the staff conducted its full operating license review. By that time changes were difficult and extremely expensive. The staff and construction team resolved the problem to some extent by maintaining informal contact through the construction period.

Because the reactors were experimental and the technology was not well understood, the AEC could not confidently assess the likelihood or the consequences of an accident, nor could it develop reliable design or siting criteria in this early phase. Instead of objective, quantitative measures, the regulatory staff and the Advisory Committee on Reactor Safeguards developed an *ad hoc, subjective* "trade-off" approach to evaluating proposed designs. The qualitative spirit of that approach was captured in the following quotation from C. Rogers McCullough, Chairman of the Advisory Committee in 1958:

> In attempting to decide for a particular reactor whether a given exclusion distance provides adequate protection for public safety, the Committee *evaluates* design features such as containment vessels, missile shields, biological shields, hydrology, meteorology, and geology, all of which affect reactor safety, particularly when a reactor is located near a populous area. Thus *it was felt* that the Elk River site would provide an *acceptable* degree of protection to the public, in view of the isolated primary system and the vapor containment provided. *Like considerations were applied* in the case of the pressurized water reactor. The Sodium Reactor Experiment had *somewhat less containment,* but has a *greater* exclusion radius than the others mentioned.[12]

The small size of the reactor allowed the agency to rest its approach to safety on the simple mechanism of containment coupled with a policy of siting away from populated areas. Reactors were surrounded by strong steel shells capable of withstanding major internal increases in pressure and of "containing" any radioactive releases that might result from a reactor malfunction. Given the uncertainties inherent in working with a new and untested technology, the Commission continued its strategy of confining the consequences of an accident rather than guaranteeing prevention.

Although the regulatory process required certain formal steps, its predominant characteristic in the early days of reactor development was informal cooperation. The applicants had little experience with the technology and often were seeking AEC support on the project, so they worked closely with the AEC reactor development and licensing staffs. Because the reactor development staff had the strongest engineering support and the greatest operating experience, the regulatory staff relied heavily on that group for information and advice on what safety requirements should be made. The development staff, eager for the successful application of the technology—especially projects it was funding—worked closely with its fellow AEC engineers down the hall in the Hazards Evaluation Branch. The applicant and the AEC staff had a common objective, the successful development and application of nuclear reactor technology.

The Commission, for its part, made licensing decisions removed from public view and without any explanatory public record. The secrecy shrouding its decision-making eventually led to what appeared to be a congressional consensus that AEC licensing decisions were almost "personal" decisions based on intuition and nonregulatory criteria.[13] That is, in the view of Congress, the chairman, Lewis Strauss, dominated the decision-making and tended to show more concern for commercializing the technology than insuring that it was safe.

In the first years of the commercial reactor program, neither Congress nor the public expressed dissatisfaction with this way of regulating reactors. The program was new and very small. Congress trusted the Joint Committee and the Joint Committee trusted the technical competence of AEC/reactor manufacturer teams. The public was hardly awakened to the existence of the program and certainly not to the fact that it might present a possible danger.

The Fermi Incident

The first safety controversy occurred in 1956, when development and regulatory interests collided. A year earlier, in March of 1955, a consortium headed by Detroit Edison and called the Power Reactor Development Corporation (PRDC) submitted a proposal to the AEC for assistance under the first round of the Power Reactor Development Program. The consortium proposed to build and operate a 100 MWe sodium-cooled, fast-breeder reactor to be called the Fermi reactor near Lagoona, Michigan, twenty miles outside Detroit.[14] The consortium's proposal came as the AEC's own Experimental Breeder Reactor, EBR-1, in Idaho suffered an accidental partial core meltdown of unknown origin.

The commissioners, Chairman Strauss in particular, gave the consortium every encouragement. They were eager that faith in the breeder concept be restored. They believed that the breeder, with its capacity to produce quantities of by-product plutonium, was a very promising technology. And if this full-scale demonstration plant could be successfully brought on line, development time might be cut by several years.[15] Moreover, the Fermi proposal for support under the Power Reactor Demonstration Program was one of the first to be made by a privately owned utility.[16]

In January of 1956, after a favorable reception by the development staff and while the contract for AEC support was pending, the PRDC submitted its application for a construction permit. The application was favorably reviewed by the Hazards Evaluation Branch staff and passed on to the Advisory Committee on Reactor Safety. The Advisory Committee, however, remembering the still unexplained EBR-1 accident, recommended against granting the permit. In its confidential report submitted to the Commission in June of 1956, the committee concluded that there was "insufficient information available at [that] time to give assurance that the PRDC reactor [could] be operated at that site without public hazard."[17] The committee went on to suggest that, as a minimal prerequisite to issuing the construction permit, the Commission conduct a detailed run of tests on the damaged EBR-1 remains and on the new EBR-2 to understand the cause of the partial meltdown and to gather additional operating data on the fast breeder design.[18]

AEC Commissioner Murray leaked work of the Advisory Committee report to the Joint Committee on Atomic Energy, although the Commission refused to release the full report to the committee.[19] The Joint Committee's public power enthusiasts, who had always supported government-owned and operated proto-type plants as a first step toward government-controlled commercial nuclear power production, quickly condemned private enterprise's first real foray into the nuclear arena on grounds of safety. This opposition notwithstanding, on August 4, 1956, the Commission issued a conditional construction permit stipulating that the Fermi plant could be built, but requiring that certain problems be resolved before an operating license could be issued.[20]

Senator Clinton Anderson (Democrat/New Mexico) and Representative Chet Holifield (Democratic/California), both supporters of public power, wasted no time in meeting this dual challenge to the committee's authority. Holifield spurred organized labor and G. Mennen Williams, Governor of Michigan, to intervene and demand a public hearing on the grounds that (1) the reactor was, if not a safety hazard, at least not demonstrated to be safe, and (2) the site, so near Detroit, was not appropriate for such a plant.[21] And in the Joint Committee's lengthy hearings on the subject, Anderson threatened to introduce legislation requiring full public disclosure of safety hazards associated with each plant and dividing the AEC into separate promotional and regulatory units.

As *Science* magazine noted several years later,

> Much of the strong feeling about the case stemmed from a conviction that the public was being put in danger in order to make it possible for private power interests to take over atomic power, a situation which, if true, would strike supporters of public power as especially outrageous

after the possibilities of atomic energy had been developed only through vast public expense.[22]

The AEC public hearing commenced in January of 1957 and ended in August of that year. To avoid further charges that its dual role as developer and regulator jeopardized its ability to decide the question fairly, the Commission created a "separate staff" to prepare its case. This staff was to act independently and not discuss the case with any member of the Commission. In November the Commission requested that the Hearing Examiner certify the entire record over to the Commission without a decision on the grounds that the examiner, an administrative lawyer, had no previous background in atomic energy matters.[23] The case was then argued before the Commission, who had no staff help because the staff was "separated." And in December of 1958 the Commission continued the PRDC's permit in force, ruling that assurances of safety need not be complete, only that the probable outcome must be favorable at the construction permit stage.[24] The intervenors took the decision all the way to the Supreme Court. They finally lost in a seven-to-two decision on the procedural grounds that (1) a construction permit did *not* necessarily entitle the applicant to an operating license, and (2) it was the operating license review that must prove that the public safety was assured.[25]

A contrary decision would almost certainly have ended the participation of private power interests in the development of nuclear power. During this evolutionary phase of reactor development, new reactors invariably included major new design features. Therefore, utilities could hardly have been expected to meet strict safety requirements at the preconstruction stage without locating so far from population centers that they could not have marketed their power.

The Fermi contest represented a long "licensing delay," typical of many that were to come in later years. But in fact there was no real delay in construction, since the permit was in force through the entire proceeding. Although PRDC did not have to pay for an actual construction delay, the consortium did have to live with and plan around the uncertainties of a prolonged ambiguous situation.

In August of 1956, when it became clear the Commission intended to issue the Fermi reactor construction permit, the Joint Committee on Atomic Energy scheduled hearings to explore ways of forcing the Commission to adopt more open, responsive procedures. The committee explored three possible procedural amendments to the Atomic Energy Act of 1954: (1) mandatory prelicensing hearings, (2) immediate publication of all safety analyses, and (3) division of the AEC into separate promotional and regulatory agencies.[26]

In its report to the committee, the committee staff gave considerable attention to the rationale for dividing the agency.[27] The report noted that the Commission's development objectives combined with industry's need to cut costs in order to be competitive could well influence the standing the Commission would give to safety requirements.[28] Most participants in the hearings generally accepted the staff argument that regulatory and development functions were inappropriately joined in the AEC. However, they were extremely reluctant to divide the agency at that moment because nuclear expertise was in very short supply and because the regulatory and development programs were necessarily intertwined at that stage in the reactor's evolution. As the Commission observed in a letter to the Joint Committee,

> We believe such a division would seriously hinder and delay [the development of a national atomic energy program] and result in waste and duplication of effort, particularly in view of the present state of the art.[29]

The alternative strategy agreed to in principle by the Commission and the committee was to separate and isolate the Hazards Evaluation Branch staff within the agency.

The second issue receiving considerable committee attention was the role of the Advisory Committee on Reactor Safety. In its early years, the Commission needed and relied heavily on the ACRS as its only source of competent safety advice. As the Hazards Evaluation Branch staff grew in size and competence, the Commission understandably came to rely on its own in-house group. As competition for project review authority grew between the two groups, the Commission argued that the Hazards Evaluation Branch staff should have review authority, requesting ACRS assistance only as needed for new and unusual plant designs or special problems. The Commission believed that the ACRS should concentrate on developing generalized design and site criteria and standards.[30] The ACRS, probably realizing meaningful standards or criteria could not be developed at this time, given the information gaps and the rapid pace of technological evolution, fought to retain its project review role.

In the context of the Fermi permit controversy, the ACRS found a congenial ally in the Joint Committee. The committee argued that there was a role for both the Hazards Evaluation Branch staff and for the ACRS and backed up its position with legislation giving the ACRS statutory existence and formal review responsibilities for each application. Specifically, the legislation provided that

> the Committee shall review safety studies and facility license applications referred . . . and shall make reports thereon, shall advise the Commission with regard to the hazards of proposed or existing reactor facilities and the adequacy of proposed reactor safety standards, and shall perform other duties as the Commission requests.[31]

It further provided that the ACRS could review reactors or problems on its own initiative and that all reports to the AEC be public documents.

In testimony before the committee, Senator Anderson noted that "the procedures [were] intended to help increase public knowledge of reactor safety problems and control, and also to help assure fair and impartial administrative actions on applications." Moreover, the committee argued that since the federal government was proposing to subsidize reactor liability insurance through the pending Price-Anderson Act, the ACRS review offered additional assurance that the reactors had been fully evaluated.[f] The private insurance companies providing the base liability coverage testified that they, too, relied heavily on the ACRS's judgments.[32]

The committee considered a second proposal to increase the formality of Commission decision-making and open it to public view: a mandatory hearing on every license application. The Commission opposed the measure, arguing it would delay action on licensing applications while offering no real extra benefits. Although the issue sparked very little debate, the committee did support the measure as a means of providing interested parties with easy access to the licensing process without forcing them to request a full hearing.

The Joint Committee incorporated the ACRS and the mandatory public hearing amendments into the Price-Anderson legislative package. Although the AEC strongly opposed the regulatory measures, since they were part of a bill providing indemnity guarantees that it very much did want, the bill got strong AEC and administration support.[33] It passed in August of 1957.

The Commission may have won its battle to issue a construction permit, but it lost much in the way of credibility and freedom of action. The Fermi case demonstrated that the AEC's subjective, personalized style of decision-making could not stand up to an outside challenge. The review process could not accommodate adversary proceedings in a way that appeared sufficiently fair or objective. And the Commission's behavior in this case cast doubt on its ability to be an impartial judge while it shouldered developmental and promotional obligations.

The well-publicized disagreement between the ACRS and the Commission also called public attention to the fact that there continued to be substantial gaps in our information about the hazards imposed by nuclear technology and that where some information existed, experts could still differ. That, in turn, suggested that rather than having the experts privately arrive at some "correct" decision, the deliberations should take on the character of an adversary proceeding: the record could expose the gaps and differences of opinion and the decision-maker could be held to that record.

Reorganization of the Regulatory Program

Although the AEC did not waiver in its stand on the Fermi reactor permit, it took hasty steps to polish its tarnishing image. In December of 1957, the

[f]The Price-Anderson Act is best known because it provided supplemental government liability insurance and set a liability limit for nuclear reactors.

Commission formally separated its development and regulatory staffs by substituting a Division of Licensing and Regulation and an Office of Industrial Development for the former Division of Civilian Applications. The new Division of Licensing and Regulation answered directly to the general manager (see figure 4-1) and was responsible for regulatory policy and procedures and for coordinating research on reactor safety. Conduct of that research, however, continued to be carried out by the Division of Reactor Development.

The Commission also made several significant procedural changes. It continued the practice of a "separated staff," where the staff reviewing an application was formally segregated and prohibited contact with the Commission and the rest of the agency to avoid conflict of interest charges. The AEC voluntarily began publishing summary staff facility analyses that included a facility description, analysis of major safety problems, and assessments of the applicant's ability to resolve them. And, as a symbol of better faith, the Commission made its initial permit ruling an announcement of a "proposed action," so construction could not begin until *after* the thirty-day hearing request period was over.

The Price-Anderson amendments took effect in 1958, requiring further procedural reforms. The applicant continued to be responsible for submitting a detailed safety analysis report, though in fact the initial reports were generally very incomplete: designs were still general and content requirements were vague, varying from reviewer to reviewer. During the protracted review process, the Hazards Evaluation Branch staff gathered necessary information and advice on the project and drafted an evaluation report for the ACRS and the Commission.

Meanwhile the ACRS began its independent review when the application was submitted. The committee's appointed subcommittee met with the applicant and the AEC staff in off-the-record sessions, contracted for any necessary outside consulting support, and finally presented its findings to the full committee. The committee then drafted its report, usually a brief two- or three-page letter to the Commission, for inclusion as part of the public record.

Once the reviews were complete, the application was prepared for public hearing. The hearing examiner, a lawyer trained in administrative law, gave public notice and held a preliminary conference to outline the issues in the case. The hearing was conducted under the rules of the Administrative Procedures Act, with the applicant responsible for preparing the case to prove his proposed plant would pose no undue risk to the public. At the conclusion of the hearing, the examiner ruled on whether the record supported issuing a permit or license, and unless the Commission reviewed the case on appeal or its own motion, the examiner's decision was final. Because the Commission generously interpreted the 1957 amendment to require full public review and hearings at *each* license stage, this process, including the ACRS review, could be repeated numerous times for a reactor, if the applicant requested temporary or provisional permits as well as standard permits and licenses.

In 1959 the Commission again reorganized its staff to further separate regulatory and development functions and personnel. It created a new post of

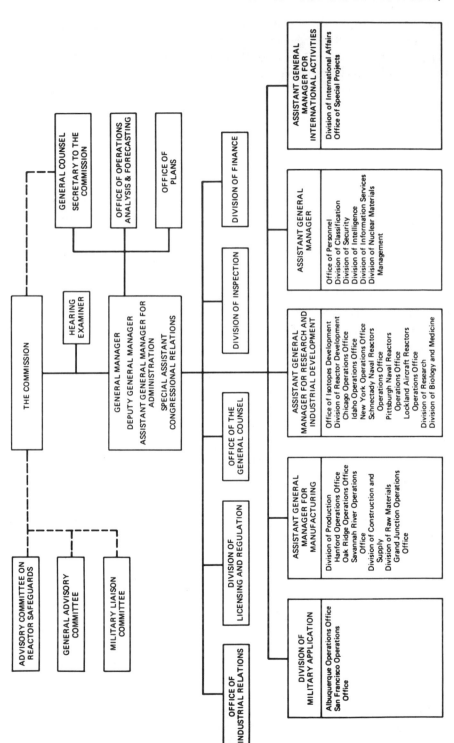

Source: *U.S. Government Organization Manual, 1959-60.*

Figure 4-1. Atomic Energy Commission Organization Chart, 1958

Assistant General Manager with independent authority over the Division of Licensing and Regulation and two new units (see figure 4-2). One, the Division of Compliance, became responsible for insuring that licensees honored the requirements of their licenses. The other, the Office of Health and Safety, was to recommend radiation exposure standards and to serve as a contact point on public health questions for states. Creation of an office specifically responsible for establishing radiation exposure standards marked a significant departure from the Commission's exclusive concern with licensing. For the first time the AEC formally acknowledged responsibility for this crucial function.

It seems likely, however, that the office was created not because of some overriding change in Commission policy, but rather in response to pressure from the Joint Committee. Nuclear weapons testing had increased ambient radiation levels considerably, causing growing public concern over the levels and over the credibility of government standards. The Joint Committee feared that the public's increasing sensitivity might spill over, creating a fear of nuclear power reactors. The committee also seemed aware that, without credible exposure standards, it would be virtually impossible to develop defensible reactor design criteria. There would always be some routine radioactive releases as well as potential accidental releases. For a reactor to be safe—and according to the definition of that time safe meant "imposing no hazard"—the releases had to be demonstrated to be without consequence.

While the principle was established, the Office of Health and Safety did little to improve safety or exposure standards. It simply adopted the existing National Committee on Radiation Protection (NCRP) standards, with slight modifications as administrative needs required. And although the NCRP was reputable, it was a private organization with no government charter and virtually no access to research funds.

To further remove itself from suspicion, the Commission also amended its own regulations to (1) bar all *ex parte* communications between applicants and Commissioners, unless they were made public: (2) require that all written communications to the Commission regarding licensing be public; and (3) make Commission-owned reactor projects subject to the same public hearings and licensing procedures that governed private reactors.

In 1961 there was a last major reorganization that consolidated the changes of the previous five years. The Commission abolished the position of Assistant General Manager for Regulation and Safety and replaced it with that of Director of Regulation. To guarantee the autonomy of the regulatory functions, the Director of Regulation reported directly to the Commission, and the three regulatory divisions—the Division of Licensing and Regulation, the Division of Compliance, and the Office of Radiation Standards—were placed under the director's control. Safety research still remained in the main body of the agency under the control of the Division of Reactor Development (see figure 4-3).

Critics of the AEC did not find the changes in organization and procedure entirely satisfying, and that same year the Joint Committee undertook yet another major investigation of the AEC's regulatory performance. Three studies,

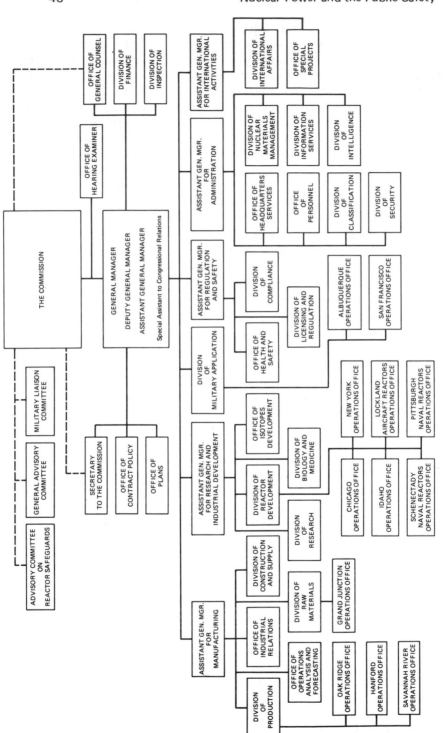

Figure 4-2. Atomic Energy Commission Organization Chart, 1960-1961

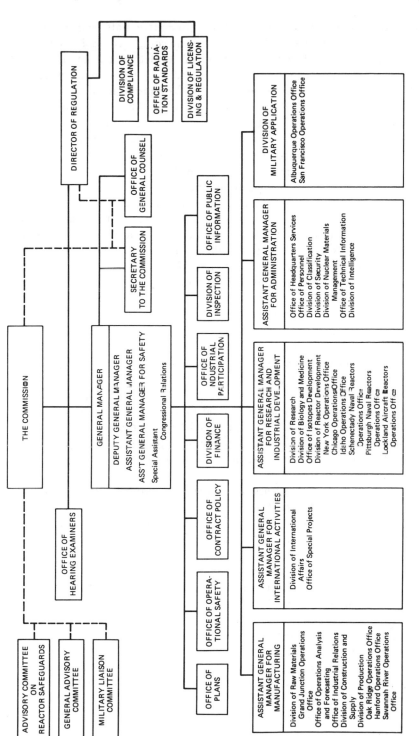

Source: *U.S. Government Organization Manual, 1961-62.*

Figure 4-3. Atomic Energy Commission Organization Chart, 1961-1962

one authored by the AEC, one by the Joint Committee staff, and one by a University of Michigan Law School study team, were completed, each reaching a different conclusion regarding the separation of regulatory and development functions. The AEC concluded that keeping the regulatory function independent but within the agency and reporting to the Commission just as it did then should satisfy those who made conflict of interest charges. The Joint Committee recommended the creation of an internal but independent licensing board. And the Michigan study argued that for reasons of efficiency and credibility the regulatory and development functions should be totally separated.[34]

A second subject of considerable concern was the hearing process. The Joint Committee intended that the hearings should provide an occasion and a forum for applicants and AEC staff to present and defend summaries of their respective positions, with clarification of assumptions and criteria. But these summaries were not intended to include a lot of technical detail.[35] The Hearing Examiner was then, on the basis of the record, to make a policy and a technical judgment as to whether there was a "reasonable" assurance of safety.

In fact, however, the hearing process fell short of the committee's expectations in several respects. First, many argued that because the permits were never contested (none had been since the Fermi application), the proceedings were not adversary and therefore were not exposing all the arguments. The record and the examiner were, therefore, inadequately informed.[36] Second, to compensate for the lack of an adversary, the examiner demanded oral testimony and engaged in lengthy cross-examination. Third, because examiners were lawyers without technical training, they were frequently unsure of the technical implications of the record and understandably seemed afraid to make decisions. Instead, they would call for reanalyses by outside experts and take interim action, calling for repeated reviews and hearings at each step or at each proposed design change.[37] Under the weight of all these pressures, the hearings became increasingly burdensome.

Not only were the hearings lengthy, but there was also considerable disquiet over the quality of the ultimate decision. An examiner trained in administrative law could be expected to guarantee a *fair* hearing to any point of view. But if the issues were technically complex, and particularly in the absence of an adversary, could he make a sound judgment on the safety of a reactor?[38] *Did the public interest demand due process or technical expertise?*

In a final effort to resolve these several dilemmas, the Joint Committee again drafted amendments for the 1954 legislation. The committee chose to support the integrity of the AEC, but to create independent licensing boards within the agency. The licensing boards were to be composed of three members: two would be technically trained and one would be qualified in administrative procedure. The committee also reduced the mandatory hearing requirements to one at the construction permit stage. Congress enacted these amendments in 1962, and from this point until the dissolution of the agency twelve years later,

the organization of the regulatory program and the program's licensing procedures underwent very little change.

Risk Analysis

During its early years, the AEC devoted very little attention to broadly
identifying or quantifying the risks associated with nuclear power production.
The Commission focused its attention exclusively on the reactor, leaving
questions of fuel cycle safety and management for a later date. In 1957,
Brookhaven Labs conducted the only risk analysis done between 1954 and
1962, and that study (Wash-740) was performed not to get information for the
regulatory program but to calculate the upper level of third-party liability as
background information for the Joint Committee's insurance program hearings.

The Brookhaven study postulated a hypothetical worst-case accident with
the maximum possible radiation release under unfavorable weather conditions
and near a high population area. It did not calculate the *likelihood* of such an
accident, only its likely consequences. The results were somewhat chilling: 3,400
dead and $7 billion in damage. The Commission became acutely concerned that
these numbers, taken out of the context of the "worst possible accident" and
used without reference to a likelihood estimate, would be misunderstood and
unduly frightening. So, rather than expanding the analysis, the AEC became
more circumspect in conducting risk estimates at all.

The AEC did not request an update of the Brookhaven study until 1965,
when the Price-Anderson insurance program came before Congress for renewal.
At that time Brookhaven prepared a second report, "WASH-740 Revision," that
showed the consequences of such an accident in the new, larger reactors to be
more severe. But estimates of the likelihood of such an accident suggested that
the probability was lower in 1965 than in 1957. The AEC withheld the
WASH-740 Revision because, as Chairman Seaborg later noted, "We didn't want
to publish it because we thought it would be misunderstood by the public."[39]
Some years later it was released when a Chicago environmentalist threatened to
sue the AEC under the Freedom of Information Act.

Regulation 1954-1962: An Assessment

Throughout these early years, both the Commission and the public seemed
comfortable with a rather murky and varying definition of "safe." In verbal
pronouncements and safety analyses, the AEC would equate Congress's "adequate protection" and "without undue risk" with "free from risk." It was
Commission policy that "there must be no 'credible possibility of an accident
that will release significant quantities of fission products' into the air."[40] In a

somewhat more vague definition of regulatory objectives, Clifford Beck, then Assistant Director of the Nuclear Facilities Safety Branch of the AEC's Division of Licensing and Regulation, told the Joint Committee that the AEC must (1) reduce the likelihood of an accident that releases radioactive materials to "the *lowest practical* level" and (2) "*minimize*" hazardous consequences.[41] Beck made no effort to further define "lowest practical" or "minimize."

At the same hearing, Walter H. Zinn, former head of the Argonne Labs and longtime AEC consultant, defined the Commission's regulatory objective with yet a bit more leeway. Zinn suggested that the goal was to "reduce the probability of serious hazard to the public to a low enough value so that the risk [was] comparable to other risks which are found acceptable in our society."[42] Zinn's definition reflected great foresight in that it acknowledged there was risk inherent in countless human activities generally considered to be safe. The job was to define the *level* of risk found acceptable by the public.

The Commission's defined objective, on the other hand, reflected the conventional wisdom of the 1950s. In those years industry was popularly viewed as a benign and desirable neighbor. Smog had hardly been invented, and virtually nothing was known of the hazards posed by the many new chemicals just coming into use. Thus the public risks imposed by a coal-fired generating plant were thought to be negligible, and it was thought that those from a nuclear plant should be commensurate.

During this period the AEC was even less successful in developing standards or criteria by which a reactor design could be judged acceptably safe than it was in defining "safe." The technology was still very primitive—hardly beyond the experimental stage—and undergoing rapid change. Large gaps still existed in the understanding of the basic principles of reactor physics, of how materials responded in reactor environments, and of how exposure to radiation affected man and his environment. These factors combined to make the setting of useful standards and objective regulatory criteria virtually impossible even if the Commission had perceived the need and had the trained manpower to do the work. As *Electrical World* observed,

> The trouble is this: Nobody—in or out of the AEC—knows enough about the hazards at this stage of the game to lay down reliable safety criteria. The Commission and the industry must feel their way while experience is gained.[43]

There was, however, considerable evidence that the Commission did not view the setting of standards as a regulatory priority. With the exception of creating an ineffective Office of Health and Safety, the Commission did not allocate dollars or manpower to the job. Nor did it organize an information-gathering capability, either internal or external, answerable to regulatory needs. C. Rogers McCullough, testifying on behalf of the ACRS in 1961, expressed great concern over the AEC's failure to use what experimental data existed for developing objective design standards, and he went on to strongly recommend that the Commission conduct more safety research and development.[44]

Since there was no usable definition of "acceptable risk" and no design standards, those responsible for licensing reactors were left to rely on their own judgment. Virtually all the information available for the safety review came from the applicant and outside AEC consultants. Applicants were encouraged or discouraged according to the *ad hoc,* subjective assessments of that information by the regulatory staff and the Commission. Because the safety assessments were subjective and qualitative, they were difficult to defend if challenged. Again, because there was no complete set of design criteria or standards to be met by an applicant, the AEC staff had to make *de novo* assessments of each application. As applications increased in number and reactors grew in complexity, the licensing process could be expected to lengthen considerably.

It should finally be noted that participation in the regulatory decision-making was restricted to a very small group of relative experts—the Joint Committee in Congress, a small group of scientists specializing in areas related to reactor design, a few reactor manufacturers, the regulatory staff, and the licensing boards. This meant that regulatory decisions were founded on a very small base that might not reflect public sentiments at all.

The Commission itself played an increasingly minor role in regulatory decisions.[g] Whether its silence stemmed from a growing disinterest in regulatory concerns or from a fear of being subjected to more conflict of interest charges is debatable.[45] In either event, since the Commission spoke for the AEC and arbitrated intraagency disputes, lack of Commission participation in regulatory decisions inevitably put regulatory objectives and safety considerations at a disadvantage.

In spite of the several deficiencies in the AEC's regulatory program, it seemed entirely adequate for the developmental period of reactor technology. During these years, there were only thirteen reactors under construction and only four of those had been completed and were generating electricity. The early reactors were small, so reasonably distant siting and containment could be relied upon for conservative safety margins. Moreover, given the temper of the 1950s, it was very unlikely that the Commission's judgment on matters of nuclear safety would be challenged. In those years, people trusted government to act in their behalf. As a California public opinion survey discovered,

> Radiation is seen as a problem area requiring adequate planning and safeguards. However a majority of the people feel that radiation dangers would be minimal *because the utilities and governmental regulatory agencies are concerned about protecting the public.*[46]

[g]Commissioner Olsen testified that the time the Commission spent on regulatory matters fell from 50 percent in the 1950s to 10 percent by 1962. See Joint Committee on Atomic Energy *Hearings on AEC Regulatory Problems,* 87th Congress, 2nd Session, April 1962.

The AEC was a particularly prestigious agency, held in high popular regard.

During those years of reactor development, there was also a strong public tendency to defer to expert and particularly to scientific judgment. Most citizens, and congressmen, for that matter, believed the reactors were "safe"—that is, presented no unusual hazard. And since he did not understand the inner workings of the reactor, the layman also believed he could not understand or intelligently contribute to decisions affecting the relative importance of reactor safety and reactor development.

The scientific community, including policy-level AEC members, in fact, also subscribed to this notion. It did little to educate the public on the potential hazards of the atom or the problems needing resolution before nuclear technology could be widely adopted. The popular notion was that if these problems were shared with the public, the public, not understanding the context and the potential for finding satisfactory solutions, would become alarmed and perhaps withdraw support. As Sherman Knapp, chairman of Connecticut Light and Power, described the problem in an address made before the American Nuclear Society in 1965,

> [Some argue] that risk must be weighed against benefit. All we as scientists can do is provide objective information on the risks and the benefits. You, the community, must read the scales. On the face of it, this seems like a reasonable approach and I believe it is generally a well-meaning one. But the more you think about it, the more you realize that in its earnest pursuit of objectivity it strips away the very perspective the public needs. It is not really advice. People get lost in the trees, which are unfamiliar to them, and cannot find their way out of the forest. This is particularly true of the scientific information presented. . . .[47]

Notes

1. R.L. Ashley, ed., *Nuclear Power Reactor Siting, Proceedings,* February 16-18, 1965, American Nuclear Society, CONF65-0201 AEC, Division of Technical Information, p. 6.

2. See Chap. 3, Sec. 25c, Atomic Energy Act of 1954.

3. See Chap. 18, ibid.

4. For a good detailed description of the political maneuvering over the public-private power issue, see Green and Rosenthal, *Government of the Atom.*

5. Joint Committee on Atomic Energy, *A Study of AEC Procedures and Organization in the Licensing of Reactor Facilities,* 1957, p. 15.

6. Ibid., p. 38.

7. Berman and Hydeman, *The Atomic Energy Commission,* p. 76.

8. Ibid.

9. Ibid., p. 54.

10. Ibid., p. 74.

11. Ibid., p. 116.

12. Letter from Advisory Committee on Reactor Standards Chairman, C. Rogers McCullough, to John A. McCone, AEC Chairman, December 15, 1958, AEC Doc. No. 842/19. Emphasis added.

13. See Green and Rosenthal, *Government of the Atom,* and Joint Committee on Atomic Energy, *Radiation Safety and Regulation,* June 1961, pp. 310 ff.

14. JCAE, *Improving the Regulatory Process,* p. 27.

15. Berman and Hydeman, *The Atomic Energy Commission,* p. 139.

16. JCAE, *Improving the Regulatory Process* p. 30.

17. Ibid., p. 28.

18. Allen, *Nuclear Reactors for Generating Electricity. . . ,* p. 50.

19. Curtis and Hogan, *Perils of the Peaceful Atom* (London: Victor Gollancz, 1970).

20. Dawson, *Nuclear Power,* p. 193.

21. See Green and Rosenthal, *Government of the Atom,* and also Curtis and Hogan, *The Perils of the Peaceful Atom,* p. 11.

22. *Science* 133 (January-June 1961), 1908.

23. JCAE, *A Study of Procedures and Organization,* October 8, 1956, Appendix 7D.

24. Berman and Hydeman, *The Atomic Energy Commission,* p. 139.

25. *Power Reactor Development Co.* v. *International Union,* 367 U.S. 396, 419 (1961).

26. Berman and Hydeman, *The Atomic Energy Commission,* p. 78.

27. See the JCAE's *A Study of AEC Procedures. . . ,* 1957.

28. Ibid., p. 30.

29. Ibid., p. 47.

30. Ibid., p. 32.

31. Dawson, *Nuclear Power,* p. 178.

32. Joint Committee on Atomic Energy, *Hearings on Governmental Indemnity and Reactor Safety,* 85th Congress, 1st Session, March 25-27, 1967, p. 7.

33. Green and Rosenthal, *Government of the Atom,* p. 137.

34. Joint Committee on Atomic Energy, *Radiation Safety and Regulation,* June 1961, p. 243.

35. David F. Cavers, "Administrative Decisionmaking in Nuclear Facilities Licensing," *University of Pennsylvania Law Review* 110 (January 1962), 354.

36. Ibid., p. 345.

37. JCAE, *Improving the Regulatory Process,* Vol. I, March 1961, p. 30 ff.

38. JCAE, *Radiation Safety and Regulation,* June 1961, pp. 259-260.

39. James G. Phillips, "Energy Report/Nader, Nuclear Industry Prepare to Battle over the Atom," *National Journal,* February 1, 1975, p. 157.

40. *Science* 133 (January-June 1961), 198.

41. JCAE, *Radiation Safety and Regulation,* June 1961.

42. Ibid., p. 13.

43. "How AEC Licensing is Evolving...," *Electrical World* 149 (April 28, 1958), 74.

44. Ibid., p. 292.

45. For a statement of the first position, see JCAE, *Radiation Safety and Regulation,* p. 28, and for the second see Berman and Hydeman, *The Atomic Energy Commission,* p. 169.

46. Internal AEC Memorandum, John A. Harris to the Commissioner, May 26, 1967. This memo reported on a survey sponsored by six California utilities and conducted by ORC-West, Inc. Emphasis added.

47. R.L. Ashley, ed., *Nuclear Power Reactor Siting,* p. 17.

5 Early Commercialization

The problem of assuring safety is even of greater importance to economic atomic power than it is to aviation. Safety requirements amount to a substantial fraction of the total cost of construction of a reactor facility, and much expensive government and industry research is being devoted to the development of new ways to assure safety by means that are less costly. [1]

—Joint Committee on Atomic Energy, 1961

In 1953 the AEC and the Duquesne Light Company announced they would jointly build a 60 MWe power reactor in Shippingport, Pennsylvania. The Shippingport reactor generated power in 1957, becoming the first commercially operating power reactor in the United States. Ten years after the Shippingport announcement, Jersey Central Power and Light Company made an equally historic announcement. After thorough consideration of competitive bids from both fossil and nuclear steam system suppliers, the utility announced it was buying a 640 MWe boiling water reactor from General Electric for its Oyster Creek plant. For the first time, a nuclear plant competed successfully against the full range of generating alternatives. Jersey Central's decision was widely heralded as a watershed in the process of commercializing nuclear energy.

For some time, the AEC and certain of the reactor manufacturers had believed that nuclear technology was sufficiently in hand to be of commercial value. The AEC launched its Power Reactor Demonstration Program partly as a vehicle for subsidizing further reactor development, but also to *demonstrate* to the utilities the viability of the technology. Westinghouse, one of the two principal manufacturers, had a strong developmental base from which to work. With prodding from Rickover, that firm had settled on the pressurized water technology (PWR) and produced reliable reactors to power the U.S. nuclear submarine fleet. Again under Rickover's close supervision, Westinghouse undertook to design and build the first full-scale nuclear powered electrical generating plant at Shippingport. In 1957, with Shippingport well underway, Westinghouse contracted with the New England utility consortium and the AEC to build a next-generation plant. Yankee (Rowe, Massachusetts), a 185 MWe reactor, improved upon many of Shippingport's design characteristics. [2]

The General Electric Company, a major supplier of conventional utility equipment, emerged as Westinghouse's only real competitor in the reactor business. [3] General Electric had an association with nuclear reactors that predated that of Westinghouse. It had operated the plutonium-producing reactors at Hanford, Washington, since the end of World War II. And while its boiling water design (BWR) was not as successful as Westinghouse's pressurized vessel design, General Electric also developed and built submarine reactors for the Naval Reactors branch. In 1957, General Electric entered the commercial power reactor market with a contract to build a 210 MWe demonstration boiling water plant for Commonwealth Edison at Dresden, Illinois.a

Once Dresden I was underway, General Electric proposed to base further development efforts upon a strategy of comprehensive and orderly increments.[4] To this end, the company launched Operation Sunrise, a program for the parallel exploration of design alternatives through a series of pilot (experimental) and demonstration (evolutionary) plants that would culminate in a commercially competitive "target plant" by the late 1960s.[5]

General Electric was, however, unable to persuade the utilities to participate in the clearly uneconomic early phases of the program. After several frustrating years of trying, the company abandoned Operation Sunrise and, skipping further development steps, entered the competitive market with its Oyster Creek bid: a turnkey contract to build a full-scale plant at a price it knew to be competitive with fossil fuel.b

While General Electric knew what price its fossil competitors commanded and designed the turnkey contract specifically to underbid them, it apparently did not intend to lose money in the long run.c What the company did expect was to win utility confidence with its competitive offer and thus gain a toehold in the market. Then, after selling three or so similar plants for small losses, it anticipated moving into the black with additional sales of the same basic 600 MWe design.[6]

Westinghouse quickly followed General Electric's lead, contracting to build two pressurized water reactors in 1965 on a turnkey basis. Between 1963 and 1967, both manufacturers contracted to build a total of twelve turnkey plants (see figure 5-1). Thereafter, that form of contract was abandoned in the United States.

By offering turnkey contracts at prices competitive with fossil plants, reactor manufacturers reflected an astonishing assurance that the nuclear technology was very well in hand. To make such an offering, they had to believe that, in spite of the fact that they were proposing a new generation of reactors in the 400-600 MWe range (a range well outside previous experience), no major design, construction, or performance problems would arise to distort their projected costs. Considering that only six commercial reactors using the light

---

aFor a brief description of the differences in these two types of design, see figure 3-1.

bA turnkey contract calls for the complete financing, construction, and testing of the specified unit for the bid price. By using the bid format, the manufacturers of nuclear steam supply systems could greatly reduce the special risks to the utility inherent in opting for new and untried technology, thereby guaranteeing that the low bid would be competitive.

cIn fact, General Electric ending up loosing $500 to $750 million on its first seven turnkey plants, a result perhaps best described as the price of premature "forced entry" into the industry. See Perry, p. 94.

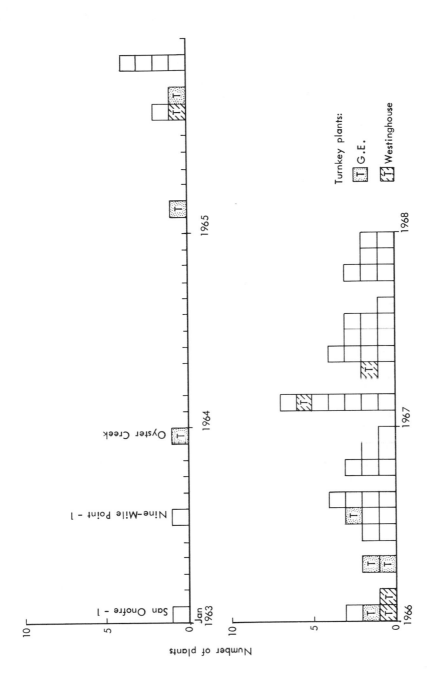

Source: A. Gandara, *Electric Utility Decisionmaking and the Nuclear Option*, p. 54. Santa Monica: The Rand Corporation, R-2148-NSF, June 1977.

Figure 5-1. Orders for Nuclear Plants, 1963-1968, by Month

water configuration had been completed and were actually generating electricity by 1965, the manufacturers were clearly making such judgments in the absence of much experience.

Nonetheless, optimism prevailed. Between 1963 and 1967, utilities contracted for fifty-nine nuclear reactors, accounting for 30 percent of the total steam electric capacity ordered over those years. Of these fifty-nine reactors, twelve were bought under turnkey contracts. But during the same period, utilities ordered forty-seven reactors under contracts with no turnkey provisions (see figure 5-1). Their willingness to agree to conventional contract provisions clearly demonstrates that the utilities, like the manufacturers, believed nuclear technology was indeed well in hand and could be relied upon to compete with fossil plants. The AEC, seeing commercialization proceed so rapidly, considered its own development function adequately fulfilled and began phasing out its demonstration program and fuel subsidies. The technology had presumably come of age.

The nuclear reactor's competitive strength lay in its low fuel costs. Between 1961 and 1967 the cost of fossil boiler fuel ranged from about 1.3 mils per kwh in Wyoming to 7.8 mils per kwh in Vermont, averaging 2.6 mils to 2.8 mils nationwide. By comparison, nuclear fuel costs ranged from 1.1 mils to 1.9 mils per kwh over the same period.[7] And during those years the AEC and the nuclear industry believed cheap radioactive waste disposal techniques could easily be developed, so no additional waste storage or disposal charges were calculated for nuclear plants. Operating and maintenance charges were roughly the same for both types of plants.[8] The reactor's competitive weakness lay primarily in its considerably higher capital costs (which include construction and financing costs). On balance, the cost analyses being done in the early 1960s showed the fossil and nuclear 600 MWe plant to be roughly competitive for areas where the cost of fossil fuels was high.

In addition to the normal costs of construction, operation, and fuel, nuclear plants were subject to unique and potentially costly regulatory requirements that did not apply to fossil plants. Siting constraints requiring generating facilities to be located away from load centers imposed two types of additional costs. First, there would be the cost of installing and maintaining transmission equipment. Second, electricity would be lost in the course of transmission. A further siting consideration was access to cooling water. Nuclear plants produced almost a third again as much waste heat as fossil plants, and hence needed easy access to a large supply of cooling water.

Changing regulatory requirements to require backfitting or new safety features could escalate capital costs substantially. And lengthening lead times and licensing delays that deferred the time when a utility began earning a return on its investment could also jeopardize the nuclear reactor's competitive position.

During the 1950s the fledgling nuclear industry kept a low profile, lobbying little and enjoying a junior partner relationship with the AEC. As nuclear technology entered the marketplace, the industry viewed the additional regulatory costs as potentially ruinous and became an outspoken advocate of an "industry position." Chief among its objectives was to persuade the AEC to relax its opposition to siting in or near metropolitan areas and thereby to make transmission costs for nuclear plants comparable to those for fossil plants. Industry's position was ably summed up in testimony by Westinghouse spokesman Joseph Rengel before the Joint Committee on Atomic Energy in 1967.

> Metropolitan sites for nuclear power reactors can be justified on economic grounds at this time. Further, we believe the plants can now be designed, built, and operated safely in metropolitan areas. At the present, the major obstacle is the lack of definitive requirements for satisfying the AEC regulatory bodies. We think these requirements can be defined now.[9]

As part of the same effort to get economically advantageous siting, utilities wanted access to the coasts, where abundant cooling water was readily available. The West Coast, however, presented special siting problems because of its high level of seismic activity.

Second, the industry pushed hard for a short licensing period with no unexpected delays. A persistent theme in industry testimony before the Joint Committee's 1967 Licensing and Regulation Hearing was that standardization was imminent and that it should lead to a major simplification in the licensing process. Advocates argued that once components or systems had been used and approved in one plant design, they should not need reevaluation in a second. Also, to make licensing requirements predictable and reduce the leverage of the AEC staff in changing plant requirements, the industry volunteered to participate in a joint effort with the AEC to draw up a set of comprehensive, firm standards and criteria to govern siting and plant design.

Third, the industry understandably wanted to preclude the imposition of additional, unexpected costs once a project was underway. Backfitting and ratcheting loomed as serious threats.[d] This was especially true after 1965, when operating experience began to give hints of design deficiencies and the increasing size of the new plants prompted increasing concerns about safety margins.

Siting

The nuclear industry was particularly concerned about the economic penalties of remote siting. Transmission, including construction and maintenance, cost a

---

[d]"Backfitting" and "ratcheting" are AEC jargon. Backfitting refers to the modification of an operating facility. Ratcheting refers to the tightening of applicable standards or requirements for a plant that is still in the design or construction phase. In the latter instance, no retrofitting is actually required, but major design and engineering changes might be required.

utility in the neighborhood of $250,000 per mile, while electricity losses from the line and increased system instability could incur similar additional costs.[10] The importance of transmission costs in a utility's choice of generating alternatives was reflected in a 1962 report by Sidney Stoller & Associates[e] that predicted that the additional costs imposed by AEC siting requirements of that date would shift the potential nuclear share of all new generating capacity ordered from 20 to 50 percent to 0 to 20 percent.[11] And, in fact, several utilities had already declined to participate in the Power Reactor Demonstration Program because siting requirements increased costs too much.

Although early in the reactor development process the Commission had adopted no explicit policy requiring very remote siting, remote siting had become accepted practice.[f] The early reactors were all AEC-owned test and plutonium production facilities. These reactors had no containment or any other safety features. But they were small, and security as well as safety dictated they be sited in regions remote enough to make them a negligible hazard. However in 1953, when the Commission debated what reactor commercialization strategy to take, it noted that "the present conservative safety standards followed by the Commission in connection with the design and location of nuclear reactors will require complete reexamination in light of the economics and practicalities of private nuclear power development."[12] Even at this early date, it was probably anticipated that a combination of more operating experience and some system of plant safeguards would allow the Commission to give up its remote siting requirements.

The containment structure became the Commission's first required safeguard. It first appeared as a single steel sphere on General Electric's Knoll Laboratory reactor in West Milton, New York and as multiple structures enclosing the reactor and the generating facilities of Shippingport. Later the Commission required all Power Reactor Demonstration Program reactors to be contained in steel-lined structures capable of withstanding the pressures expected from a large primary system rupture. But containment was not openly viewed as a substitute for distance, only as a complement to it.

In 1959 the Commission issued proposed siting guides formally specifying required exclusion and population center (defined as 25,000 or more people) distances based on the reactor's designed power output. The nuclear industry vehemently opposed the guides, arguing that they were unnecessarily conservative.

The Commission modified the guides and reissued them two years later, by which time a major segment of the industry had come to prefer conservative guides to no guides at all.[13] In addition to the traditional exclusion area and

---

[e]Sidney Stoller & Associates is one of the most reputable consulting firms in the nuclear business. Their conclusions were certain to be heeded.

[f]In the late 1940s the AEC seemed to be using a formula that required the order of 5.5 miles (an area of 15,000 acres) for a 100 MWe plant. Perry and others, *Development and Commercialization...*, p. 103.

population center criteria, the new guides required that reactors be somewhat isolated from low population areas as well. While the guides imposed specific, quantitative siting requirements, the Commission softened their impact substantially by affirming that they could be violated "if the design of the facility [included] appropriate and adequately compensating engineered safeguards."[14]

In 1962, the Commission formally adopted the guides as its Reactor Site Criteria, more commonly referred to by their *Federal Register* title as 10CFR100 (see following). Since they were criteria, the provisions of 10CFR100 offered performance standards against which proposed reactor designs could be measured, not definitive specifications or design requirements. To further clarify how the criteria would be applied, the regulatory staff published a Technical Information Document, TID 14844, describing with examples how distance would be calculated. For a standard 600 MWe plant with no safeguards except containment, for instance, the exclusion area radius would be 1.9 miles, the low population area radius 28 miles, and the distance to a population center 36 miles.[15]

The economic trade-offs between remote siting and installing the engineered safeguards available at that time appeared to be obvious. In estimates that were widely circulated during the early 1960s, Harold Vann, vice-president of Jackson & Moreland, put the costs of siting a reactor twenty miles from a load center at about $12 per kilowatt capacity. He computed the costs for the full complement of safeguards at about $10 per installed kilowatt in a standard 600 MWe reactor.[16] Therefore, by using safeguards instead of remote siting, a utility could save $1.2 million on a standard 600 MWe reactor.

Manufacturers and utilities missed no opportunity to argue that safety evaluations should be based on a *combination* of site location and engineered safeguards, not on distance alone.[17] Given the economics, their concerns were quite real. Any utility big enough to absorb a nuclear plant into its system served a metropolitan area and needed a local site. Moreover, in the populous Northeast (the area of greatest market potential), if the distance formulae were honored, most sites would be ruled out entirely.[18] As Westinghouse's John Simpson noted in testimony before the Joint Committee,

> If the true technical and economic advantages of nuclear power are to be realized . . . [plants must] be located in load centers. . . . In our opinion, industry has developed reactor systems and engineered safeguards that should permit the location of large nuclear stations in population centers.[19]

By 1963, Clifford Beck, Deputy Director of Regulation, in his article "Engineering Out the Distance Factor," officially acknowledged that the regulatory staff believed engineered safeguards were sufficiently advanced to substitute for some portion of the distance requirements and would license accordingly.[20]

Substituting engineered safeguards for remote siting raised new and somewhat knotty questions. How should a safeguard be balanced against distance? That is, how many miles, for instance, should a pressure suppression pool buy? How should the effectiveness of an engineered safeguard be judged? What standards of proof should be required to demonstrate the safeguard's capability and its reliability? If the information is not available, who then should provide it? While not addressed explicitly, these questions arose and had to be answered in the course of regular Commission business.

The utilities were primarily interested in knowing what they had to do to locate at the load center. Since no standards defined an acceptable plant design for an urban site, their strategy was to submit an application for an urban site as a means of determining what kind of distance credit they could get for the available array of engineered safeguards. But in spite of continued pressure during the early 1960s, the utilities were unsuccessful in gaining access to truly urban sites.

In 1962, Consolidated Edison applied for a construction permit to begin work on its proposed 1,000 MWe Ravenswood reactor in Queens, New York. The AEC and the Advisory Committee on Reactor Safeguards did not reject the proposal out of hand, but they viewed it with obvious skepticism, and eventually Consolidated Edison announced it was canceling the plant.g A year later, the Los Angeles Department of Water and Power requested a construction permit for its proposed 492 MWe reactor at Malibu. If the application had not been killed because of seismic problems (see below), it probably would have faced a stiff challenge on grounds of urban proximity. One last attempt was made in 1966 by a utility consortium called Public Service Electric and Gas Co. which proposed a 993 MWe reactor for a site near Burlington, New Jersey. Again both the AEC and the Advisory Committee on Reactor Safeguards informally indicated strong reservations, and the site was shifted to Salem, New Jersey.

During the 1967 Joint Committee Hearings, Representative Craig Hosmer (Democrat/California), a key member of the committee, inquired of Harold Price, Director of Regulation, "Can you give any specifics about what is needed before the AEC would allow a reactor to be sited in a metropolitan area?" Price's answer was simply, "No sir." He went on to point out that good research data, operating experience, and testing and inspection capabilities were probably minimum prerequisites, and that the AEC had access to none of those capabilities at that time. The Advisory Committee on Reactor Safeguards concurred in the AEC's reservations.[21]

Although no urban sites were approved, the regulatory staff was clearly willing to give some credit for engineered safeguards. During the same year the siting criteria were adopted, Southern California Edison's 400 MWe San Onofre reactor was granted a construction permit. In return for installing a safety

---

gConsolidated Edison claimed it canceled the application because it unexpectedly got access to some extra Canadian power. But even if that was the real reason for cancellation, all the evidence suggests the application would have been denied if pressed.

injection system of borated water designed to fill the reactor vessel and hold expected fission product releases to 6 percent of their normal value, Southern California Edison was able to reduce San Onofre's exclusion zone from .82 miles to .5 miles and its distance from a major population center from 12.5 to 4 miles. After that, reductions became the rule.[22]

The array of possible safeguards fall into four categories: (1) those that contain the fission products during and after an accident, preventing leakage to the outside, (2) those that reduce the pressure from an accident sufficiently to let the containment work, (3) those that reduce the fission products that *can* leak (filters, for example), and (4) for small accidents, those that reduce the effects of a release (tall stacks, dilution devices, and so on).[23]

By the mid-1960s reactor designs had incorporated examples, albeit sometimes modest, of all four categories.[h] But only a few of these reactors had been completed and none had put any safeguards through the test of a real accident. So apart from laboratory scale experiments and engineering calculations, there was little evidence to demonstrate exactly what the capabilities of these safeguards were or with what reliability they would function. Moreover, lack of operating experience and the generally poor understanding of exactly what course a serious accident might take meant that the uncertainties were compounded.

To compensate for the uncertainties, members of the regulatory staff gave only partial performance credit to safeguards in their accident release calculations, and where possible they also required redundant safeguards.[24] How much credit to give and when redundancy should be required were totally subjective decisions left to the discretion of the project staff and subject to the review of the Advisory Committee on Reactor Safety and the Atomic Safety and Licensing Boards.

Applications to site reactors on California's seismically active coast presented the regulatory staff with similarly difficult decisions. In an area where water was particularly scarce, coastal siting appeared to be an economic imperative. But there was considerable uncertainty about how to identify an active fault area, about what kinds of stresses a quake actually exerted, and about what given structural designs could actually protect against. Probably because the proposed coastal sites were somewhat removed from population centers, the regulatory staff and the Advisory Committee on Reactor Safeguards responded with more ambivalence to the possibility of siting in seismic areas. But because the licensing progressed further, these cases also provided the opportunity for the first rumblings of citizen opposition to nuclear power.

In 1962, Pacific Gas and Electric (PG&E) applied for a construction permit

_____

[h]These examples included reinforced containment structures, containment cooling systems, pressure suppression devices, chemicals and filters to precipitate out or hold up releasable fission products, and a primitive emergency core cooling system to provide make-up water in case the reactor lost its regular cooling water. See Perry and others, *Development and Commercialization . . .* , pp. 103-104.

for a 325 MWe boiling water reactor at Bodega Head, a site fifty miles north of San Francisco and very near the San Andreas Fault. The staff argued that while the PG&E design was very likely to be able to withstand any expected quake activity, its adequacy could not be empirically demonstrated. In its report of October 26, 1964, the staff therefore concluded that the site was not suitable and that a large reactor should not be the subject of a "pioneering construction effort based on unverified engineering principles, however sound they may be."[25] The staff went on to recommend that the permit be denied. The Advisory Committee on Reactor Safeguards, on the other hand, believed that the engineering principles and general design for the proposed plant suggested it could withstand the forces expected from a quake in the fault zone. The committee was therefore sufficiently assured it could be operated "without undue hazard" to recommend that the commissioner approve PG&E's permit application.[26]

While the Advisory Committee and the regulatory staff were trying to resolve their differences, PG&E encountered unexpected opposition from a totally new quarter, a local citizens' group. The Bodega Head Association, a small group organized to oppose the construction permit, initially objected to the plant because the Association opposed development of the unique coastal area. Only later, as its support and the intensity of its campaign grew, did the association add the seismic characteristics of the site to its list of concerns. Surprised by such intense local opposition, PG&E finally abandoned its plans for Bodega Head in November of 1964 and agreed to work jointly with the Sierra Club to find an acceptable alternative coastal site.

A year after PG&E applied for its construction permit, the Los Angeles Department of Water and Power submitted its application for the Malibu plant, also to be located on the coast in an active seismic area. In this case the Advisory Committee on Reactor Safeguards and the regulatory staff agreed that the design was adequate and that a construction permit should be issued. But once again the utility encountered unexpectedly stiff opposition on the seismic issue from the community. Five intervenors contested the permit, including comedian Bob Hope (a participant repeatedly noted in the AEC references to the case, therefore, presumably one to be reckoned with).

In 1966, after protracted hearings, the Atomic Safety and Licensing Board handed down a split ruling allowing the Los Angeles Department of Water and Power to proceed on the basis of a provisional construction permit but also requiring that it strengthen the seismic characteristics of the plant design. In 1967, both the regulatory staff and the intervenors appealed the ruling to the Commission. The Commission's final decision supported the intervenor's contention that the applicant's design had to meet more stringent specifications and then be reevaluated before a permit could be issued and construction begun. The Los Angeles Department of Water and Power then withdrew its application and abandoned the project.

During the same period, Southern California Edison received a construction permit for its 400 MWe San Onofre reactor to be built on the coast next to San Clemente after only minor intervention and an AEC study of the surrounding geology. Then, in 1967, PG&E applied for and, after a contested hearing, was granted a construction permit for Diablo Canyon, its 965 MWe plant on the coast west of San Luis Obispo. At both San Onofre and Diablo Canyon there was clear evidence of seismic activity in the vicinity, though neither location was thought to be in a major fault area.

Thus the regulatory response to the uncertain hazards of siting reactors near seismic areas appeared to be very similar to that of siting in urban areas. While the various regulatory decision groups (the regulatory staff, the Atomic Safety and Licensing Boards, the Commission, and the ACRS) were sympathetic to economic necessities and never suggested that commercial use of nuclear power should be discontinued until the uncertainties were resolved, their decisions reflected the very real technical uncertainties and a perceived need for a certain degree of conservatism. Their assessments were also clearly subjective and qualitative.

Standards

As a prelude to standardizing their plant designs, reactor manufacturers began pressing the AEC for a set of design standards and more detailed siting guides in the early 1960s. They argued that sound standards insuring good design and construction would contribute considerably more to public safety than would the "procedural niceties" of hearings and public decision-making.[27] The Joint Committee, fearing that the vagaries of a licensing process based on subjective, personal judgments might unduly discourage commercialization, gave the manufacturers full support.[28]

But asking for design standards was asking the impossible. The same kinds of uncertainties that precluded clear and quantitative siting standards prevented the AEC from developing a useful body of general design and component standards. Plant design was changing rapidly, requiring the constant reevaluation of new systems and interrelationships. And even by 1966 only six commercial plants were in operation, all 265 MWe or less. Thus, there was precious little operating experience upon which to base standards, and none of it from the type or size of plant then being proposed. Without operating experience or well-tailored research data, the regulatory staff had no basis for evaluating failure rates and predicting reliability. It could only extrapolate from current, noncomparable data. Nor could the staff be certain of the long-term effects of radiation on materials. In spite of the obvious obstacles, for political reasons the AEC had little option but to try, all the while arguing the standards should be general and flexible.

To accommodate the external realities, the regulatory staff evolved four types of standards.[i] *Standards* were definitive requirements specifying procedures for obtaining and confirming related performance requirements. *Codes* constituted special types of standards, detailing design requirements for components or systems. Codes were generally drawn up by the relevant professional societies and were rigorously applied and enforced.[j] Then, to handle more ambiguous situations, the staff came to rely on two other measures: *criteria,* which were performance objectives or yardsticks against which to compare the applicant's proposed design, and *guides,* which were "suggested" ways of dealing with "possible" safety issues. Criteria and guides reflected standards that were clearly acceptable to the regulatory staff, and therefore designs observing them could expect less scrutiny in the review process. They were not, however, inviolate.

Standards, codes, criteria, and guides were all codified under Title 10 of the Code of Federal Regulations. 10CFR20 specified general radiation protection standards. 10CFR50 spelled out licensing procedures and plant design requirements. And, as noted above, 10CFR100 dealt with site considerations and requirements.

Because it did not begin to have the information necessary for drafting meaningful design standards, the regulatory staff produced a series of very general criteria including the General Design Criteria for Nuclear Power Plants, a revised Guide for Organization and Contents of Safety Analysis Reports, and the Reactor Site Criteria. At the same time, the staff contracted with MIT to amass and synthesize all research findings that might provide data for setting reactor standards.[29]

The staff also pushed the professional societies and the American National Standards Institute to develop new codes to accommodate the unique demands of nuclear plants. The traditional industry-based codes were not adequate, since no other technology subjected materials to the same kinds of stress, required the same degree of leak-tightness, or demanded the same level of reliability that a reactor did. The regulatory staff was particularly eager to get a new section for nuclear pressure vessels drawn up by the American Society of Mechanical Engineers for its Boiler and Pressure Vessel Code. There were no codes for nuclear reactor components although the pressure vessel served as the foremost barrier to escaping fission products. And pressure vessel failure was arbitrarily considered an "incredible" accident. Therefore no measures were required to control for the consequences of such an event.

---

[i]The rules and regulations adopted by the Commission can be found under Title 10 of the Code of Federal Regulations and result from the authority conferred by Section 161 of the Atomic Energy Act of 1954.

[j]American professional societies have traditionally drawn up quality assurance specifications for products related to their areas of expertise. For instance, the American Society of Mechanical Engineers has a Boiler and Pressure Vessel Code that antedates the nuclear reactor by a number of years.

Although the regulatory staff made a serious effort, little really changed by 1967. Dr. A. Eugene Schubert, general manager for General Electric's Nuclear Energy Division, complained that

> as a system today . . . this historic case-by-case, *de novo* review procedure presents the industry with troublesome uncertainties—both technical and economic. Specific site considerations aside—neither manufacturers nor utilities can be fully confident that a reactor that was licensed yesterday can be licensed today. This makes for difficulty in schedules and estimates of costs.[30]

Clifford Beck, Deputy Director of Regulation, countered with only mild optimism, noting that

> in the past, reactors have been evaluated and particular safety requirements specified more or less on a case-by-case basis, with regulations, guides and instructions setting forth mostly the principle areas of consideration and the generally recognized performance requirements. More recently, as reactor technology has improved and as water reactor designs have tended toward standardization, the possibility of developing more specific safety criteria has appeared. . . .[31]

And finally, Nunzio J. Palladino, speaking for the ACRS, argued that useful standards could not be agreed to until the rate of technological change slowed and the regulatory staff gained empirical experience.[32] Neither eventuality seemed imminent.

In addition to uncertainties, the AEC's failure to develop standards caused inequities and delays in licensing. How existing criteria were applied to a plant design depended to some degree on the staff member in charge of the evaluation. For their part, utilities, hoping to avoid detailed scrutiny of their designs, submitted very general and incomplete applications, and a long period of further inquiry inevitably ensued. Moreover, because objective standards were lacking and responsibility for assessing plant safety fell wholly to the individual regulatory staff member in charge, it is likely he made his decisions conservatively, cautiously, and only after extensive review.

Safety Research

Those responsible for licensing the reactors were obviously feeling their way during the early years of commercialization. They had to make day-to-day decisions in the presence of great uncertainties. Although the causes and courses of accidents and the burdens they might impose on plant safety mechanisms were not well understood, sites and designs had to be evaluated nonetheless. The capabilities and reliabilities of various engineered safeguards had to be assessed in the absence of operating experience. Similar difficulties afflicted various reactor systems and components.

Since uncertainties were high and decisions were necessarily based on subjective assessments, the various review bodies had their disagreements. Some were internal disagreements. But the Commission also came under attack from the nuclear industry for being too conservative and from some local citizens' groups for not being conservative enough.

All these factors suggested that the regulatory process stood in need of a strong R&D capability. A strong program, tailored to regulatory needs and begun early in the commercialization phase, could have: (1) allowed the AEC to develop standards without relying entirely on industry and the professional societies (whose membership came primarily from industry), (2) given the agency a source of credible, empirical data for use in its own regulatory program as well as in efforts to persuade the public that its standards were valid, and (3) given the AEC an opportunity to seek out potential safety problems instead of passively awaiting their appearance in on-line reactors, where they are difficult to correct. In short, such a program could have been the springboard for the aggressive and independent control of the rapidly diffusing technology.

During the 1950s, agency policymakers gave little attention to safety research activities. In the first years of the decade, safety research and reactor development research were by and large indistinguishable. Both safety and development depended on understanding how the reactor worked. Hence there was no explicit safety research program. Instead AEC scientists were generally left to pursue their own basic research interests within the constraints of their budgets. The results of their basic research were expected to lead to more efficient *and* safer reactors. By 1955, however, the seeds of an identifiable safety research program were germinating.[33] Research seemed to be focusing on reactivity excursions, boiling water reactor designs, and containment.[34] The Commission committed substantial resources to the construction of several test reactors at the AEC's National Reactor Testing Station in Idaho and contracted with outside sources for additional safety research. And in the late 1950s it launched a formal safety research program.

As the nuclear reactor stood on the threshold of commercial acceptance in the early 1960s and the nuclear industry intensified its pressure for less conservative siting requirements, the Commission sought to refocus its safety research program. Both industry and the AEC firmly believed that reactor safety requirements were too conservative. But, in the absence of more empirical evidence, the AEC's regulatory staff was very reluctant to relax them appreciably. The AEC's safety R&D program, still under the thumb of the Division of Reactor Development, set about to produce the needed information.

Frank Pittman, director of the division, chose as his priority objective defining a maximum credible accident and empirically demonstrating that its consequences could be contained.[35] At this time the maximum credible accident was defined as a single break in a primary coolant pipe. Such an event would inevitably lead to loss of coolant for the reactor core and possibly to a meltdown of the core and breach of containment. While severe, the "maximum credible accident" did not represent the worst possible case. The consequences of a vessel failure or a control rod failure leading to a nuclear excursion, for instance, would be far worse. But anything worse than a pipebreak was considered to be so unlikely that it was classed "incredible," which, in turn, meant that there was—in the AEC's judgment—no need to take precautions against such an event or its consequences.

A license applicant's safety analysis report had to demonstrate conclusively that the applicant's proposed design could withstand a "maximum credible accident" without releasing any fission products to the environment. To satisfy this requirement during the 1950s safety analysis reports simply postulated the AEC's specific maximum credible accident and guessed at its consequences given the applicant's plant and location.[36] As Commissioner Wilson told the Joint Committee in 1963, "The maximum credible accident which the Reactor Safeguards Committee [ACRS] has used as a concept, is something we have never deliberately gone out to duplicate."[37] But by duplicating the accident, he reasoned, the Commission could prove that it understood the phenomenon and could protect against it. Then the siting constraints could be reduced.[38]

On the strength of this argument, the Joint Committee supported the Commission's request for an immediate additional $20 million to begin work on several full-scale test facilities as part of a major Safety Test Engineering Program (STEP). Under STEP, the AEC began construction on a Loss of Fluid Test Facility in 1963 and on a Power Burst Facility in 1964.

The AEC designed the Loss of Fluid Test Facility to determine the consequences of its postulated "maximum credible accident" (a single break in a primary coolant line) and to assess experimentally how well the containment shell functioned under severe accident conditions, probably a core meltdown. The Power Burst Facility was to accommodate medium-scale subassembly core meltdown experiments to study fission product releases, metal-water reactions, and pressure generation under accident conditions.[39] The Commission explicitly promised that the tests would provide the large-scale experimental data that could conclusively prove the safety of the technology to the utilities and to the public and lead to a relaxation of site and design criteria.[40]

The pervasive feeling that, once the full-scale STEP tests were conducted, the reactor would be a proven good neighbor led the Commission to neglect other very important aspects of a well-rounded research program keyed to regulatory requirements. The regulatory staff needed to know for a variety of accident conditions what the fission products would be, what releases from the fuel to expect and in what form, how mobile the releases would be, and how toxic they were.[41] There was only a modest amount of good experimental data on any of these questions.

The staff also needed sound, credible data on the capabilities and reliabilities of the engineered safeguards coming into use. In November of 1964, the ACRS assessed the acceptability of the various safeguards being proposed and recommended that the Commission pursue research that would resolve the substantial existing uncertainties regarding their effectiveness.[42] Then full credit for their use could be given.

The regulatory staff needed information upon which it could base design standards, and they especially needed evidence on the effects of irradiation of materials, a very poorly understood subject in the early 1960s.[43] Finally, with the benefit of hindsight, it is also clear that, as the reactor came of age commercially, the problems of waste disposal and reprocessing warranted a major research effort.

The safety R&D program included very little work on any of these questions. The Division on Reactor Development and Technology supported some small-scale experiments at Oak Ridge National Labs and at the Transient Reactor Test facility on fission products released from fuel. It also sponsored some studies on irradiation effects and on accident probabilities and pathways.[44] But beyond that minimal effort, the safety R&D program failed to provide the information that was basic to the development of objective measures of risk.

There can be little doubt that the safety R&D program was unresponsive to regulatory needs in large measure because the Division of Reactor Development and Technology controlled the program. And since the regulatory staff was both geographically and organizationally isolated from the development staff and because the Commission itself, as a matter of choice, had declined to become interested in regulatory matters, there was no indirect way to force a more responsive program.

Also, when the light water reactor began to enjoy commercial success, the AEC came under strong pressure from the Bureau of the Budget to cut back its research in this area. Because the safety and development programs were managed by the same division, the Bureau of the Budget did not really distinguish between the two.

Licensing

Both the AEC and the Joint Committee on Atomic Energy thought the licensing review procedures that evolved during the late 1950s and early 1960s could accommodate the commercial use of nuclear power, and in fact the procedures did suffice through 1966. But it was clear by that time that, given the trends in orders and plant size, they would soon be unequal to the task.

Between 1962 and 1966, the AEC received construction permit applications for twenty-six units. Fifteen of them were submitted in 1966 alone. Over those same four years, the applicants' average reactor capacity jumped from 325 to 780 MWe, and plants increased markedly in technical complexity.[k] To handle both the growing number of applications and the increasingly complex safety analysis reviews, another twenty-three technical members were added to the eighteen-man licensing staff[l] (see figure 5-2). Staff augmentation notwithstanding, the agency could not reduce the review time (application to issuance of the construction permit) from an average of about nine and one-half months to its target of seven months. Clearly, as the new reactor orders (eight in 1965, twenty-one in 1966, and twenty-seven in 1967) became license applications, they would overload the already straining system, and the nine and one-half month average review period could be expected to increase substantially.

In hopes of forestalling such an unhappy development, in 1965 the Commission appointed a seven-member outside review panel, headed by William Mitchell, former counsel to the Commission. The Commission asked the panel to examine the regulatory program and recommend ways to expedite the licensing of reactors.[m] The Mitchell panel identified four major problems: (1) the length of the review process; (2) an absence of well-defined safety criteria and requirements; (3) increasing regulatory manpower requirements, and (4) duplication of review and evaluations. As remedies, it proposed that the AEC (1) get on with the job of developing criteria and standards, but limit its specifications to genuine safety needs, not general design specifications; (2) coordinate its safety research program with regulatory requirements; (3) define more precisely and realistically the information the applicant should include in a license application; and (4) give the AEC staff primary safety review responsibility by limiting the Atomic Safety and Licensing Board's review to the *adequacy* of the Safety Evaluation Report (precluding *de novo* safety evaluations) and by eliminating the *mandatory* Advisory Committee on Reactor Safeguards review of each application.

Ironically, given the history of the next six years, the panel noted that the hearings themselves provided an excellent opportunity to impress the public with the diligence of the AEC and the Advisory Committee on Reactor Safeguards and with the character and competence of the applicant. The panel further suggested that the licensing hearings provided the public with a good forum for expressing its views and therefore should not be abandoned in the

---

[k]The 325 MWe average omits consideration of the 1,000 MWe Ravenswood application. Also, these numbers differ from those noted earlier because they apply to construction permit applications, not plant orders. There is some lag between the two.

[l]In 1967, the regulatory division boasted a staff of 359, only 100 of whom were technically trained; sixty in licensing, fifteen in the Office of Health and Safety, and twenty-five in the Division of Compliance. The remaining 259 were administrative personnel. JCAE, *Hearings on Licensing and Regulation...*, 1967, p. 47.

[m]The Commission also hoped that the Mitchell panel would suggest ways to avoid conflicts between the regulatory staff and ACRS recommendations of the kind experienced over PG&E's Bodega Head application. While the panel did recommend one possibility, the problem was solved by Price, director of the staff, who ordered that no staff report that was in conflict with an ACRS report be issued (see following).

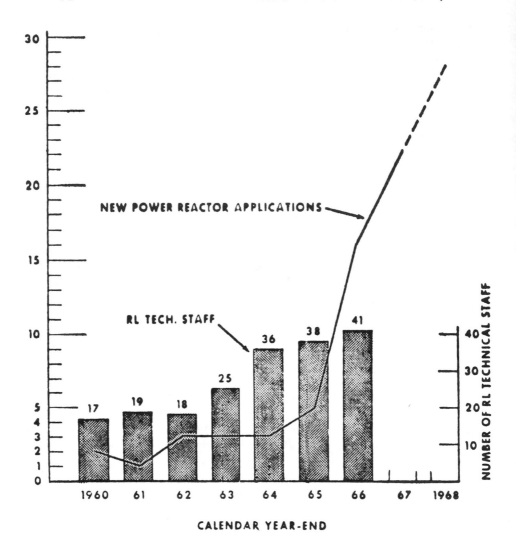

Source: *JCAE Hearings on Licensing and Regulations,* 1967, p. 407.
Figure 5-2. Increase in Power Reactor Applications Relative to the Increase in Reactor
Licensing Technical Staff

interests of faster licensing. There was, as yet, no expectation the hearings would
be transformed from a public relations opportunity for the AEC to a source of
publicity for the environmentalists' position.

The AEC made a genuine effort to comply with the Mitchell panel's
recommendations. A steering committee was formed to coordinate safety R&D
(see below). The Commission instructed the Atomic Safety and Licensing Board

to consider only the completeness of the application and review, not the validity of the conclusions, and ruled that the board's decisions would become provisionally effective immediately. The Commission also drafted and adopted a revised set of General Design Criteria to guide the applicant "in establishing the principal design criteria for a nuclear power plant."

Backfitting and Ratcheting

Throughout these early years of commercialization, representatives of the nuclear industry vociferously complained that they were continually harassed by changing agency requirements that demanded design or plant modifications. Such modifications were generally both expensive and caused delays.

In fact, the regulatory staff did frequently set new requirements after a plant's construction permit had been issued, but during its detailed design and construction phase. However, most often these new regulations did not reflect actual changes in regulatory standards. Rather they occurred because the original design approved for the construction permit was extremely general. Only after the utility moved to the detailed design phase and drew up clear plant specifications could the regulatory staff judge the acceptability of the design and its safety characteristics. As the detailed design emerged, deficiencies came to light.

Given the lack of objective standards and the constantly evolving technology, a certain amount of intervention by the regulatory staff during the detailed design and construction phase was inevitable. But the utilities themselves exacerbated the problem by keeping their construction permit applications as general and devoid of controversial issues as they could to facilitate the review process. These issues then had to be resolved in the next phase.

By the late 1960s, some plants already in operation began to experience failure in design or materials (for example, heavy vibrations, metal fractures, and so on). Combined with the growing concern over the adequacy of certain safety systems, these failures made the nuclear industry increasingly apprehensive that the AEC might impose new, stricter criteria and apply them retroactively. Utilities would then be faced with the costly job of modifying plant designs late in or after the construction phase. And, in the industry's view, a precedent for backfitting only invited a continuing escalation of regulatory requirements: the AEC would impose an endless round of new design requirements through the life of all reactors.

Although the utilities and manufacturers constantly spoke of the hardships backfitting requirements posed, they were rarely ordered in these early years of commercial nuclear plants.[n] The AEC, while not relinquishing the right to

[n]JCAE, *Hearings on Licensing and Regulations,* 1967, p. 274. Milton Shaw, Director of the Division of Reactor Development and Technology, testified that the backfitting that had been required resulted mainly from the applicant's "waiting and seeing" if a weakness in a design would go unnoticed or unchecked.

require backfitting, spoke of making backfitting requirements on a case-by-case basis and only after examining alternatives.[4 5] No one referred to the difficulty of making binding safety assessments in the absence of experimental data and operating experience.

Regulation 1963-1965: An Assessment

The main actors in the regulatory process changed markedly during the early 1960s. Having finally defined the licensing procedures to its satisfaction, the Joint Committee on Atomic Energy began to play a much less active role in the regulatory process. Conversely, the nuclear industry became an extremely active participant.

During the 1950s, few utilities showed any real interest in nuclear technology. And the manufacturers, recognizing that the technology was clearly in its development phase and the uncertainties were high, seemed content to follow rather than lead in establishing safety criteria. But during those years commercialization appeared to be a long way off.

When the commercialization process suddenly and unexpectedly gained momentum, the manufacturers became actively concerned with regulatory policies that threatened the "almost competitive" position of the nuclear reactor. Utilities, looking to nuclear power with increasing interest, joined the manufacturers in their efforts to hold down the costs imposed by regulatory requirements.

The relative strengths of the four evaluating groups, the Commission, the regulatory staff, the Advisory Committee on Reactor Safeguards, and the Atomic Safety and Licensing Boards, also shifted somewhat. The newly organized and technically competent licensing boards achieved a measure of independent authority not previously enjoyed by the hearing examiner. The regulatory staff, upgraded and expanded to meet its heavier workload, grew in stature and influence relative to the Advisory Committee on Reactor Safety. However, the open disagreement between the regulatory staff and the ACRS over granting a permit to PG&E for its Bodega Head plant led Harold Price, the Director of the Regulatory Division, to require that his group's public staff reports be consistent with ACRS findings. It is likely that Price thought that disagreements between two such expert groups might cast doubt on the competence of expert judgment and ultimately jeopardize the credibility of regulatory decisions. He therefore instructed his own staff to resolve any differences privately, and, where that was not possible, to adopt the ACRS position.

Of the four, the Commission's role shifted the most dramatically. Glen Seaborg, noted chemist and Nobel laureate, was appointed chairman of the Atomic Energy Commission in 1961. As a scientist, Seaborg had a very clear commitment to the rapid development and application of nuclear power, while his support for a strong regulatory program was ambiguous, at best. He devoted most of his energies to overseeing the agency's $2.5 billion development budget.

Commissioner James Ramey, former JCAE staff chief, appointed to the Commission in 1963, took prime responsibility for regulatory matters. He also took major responsibility for reactor development and worked closely with the Division of Reactor Development and Technology, suggesting a divided loyalty. Commissioner Ramey tended to concentrate on regulatory procedures, particularly as they applied to staff review and licensing, and showed little interest in the structure or support services required for an authoritative, independent regulatory program. The other commissioners rarely concerned themselves with regulatory questions at all.

The Commission participated in only one licensing decision during these years, and then only on appeal. In that case, it ruled in favor of the intervenors, requiring the Los Angeles Department of Water and Power to improve its Malibu plant design *before* getting a construction permit. But, on that scant evidence, one cannot conclude that the Commission tended to be more or less conservative in its safety judgments than the other review groups.

But it does appear that, consistent with their post-1957 policy, the commissioners continued to remain aloof from regulatory matters for several reasons. They were anxious to avoid the conflict of interest charges that might cause the agency to be divided. Their development mandate was far greater and more interesting to them than safety concerns, so they focused on that. And they tended to view the regulatory function as a simple legal and administrative job of conferring permits and licenses. Physically, the regulatory staff was relocated to Germantown, Maryland, while Commission headquarters remained in Washington. The physical separation led to a loose structure of absentee management with most regulatory decisions resting with the regulatory division director.

Not only did the roles of the actors change, but a new participant entered the stage during these years of early commercialization: the intervenors. The nuclear industry and the regulatory staff saw the intervenors primarily as a product of poor public relations. That is, rather than seeing in them the reflections of legitimate public concern and changing public values that demanded substantive response, they saw technically uninformed people who needed to be taught how risk-free and beneficial nuclear power indeed was.[46] They proposed to change public attitudes through better information programs and such cosmetic tricks as relabeling the *Hazards Evaluation Report* the *Safety Evaluation Report*.

In spite of the consternation the intervenors provoked in the Bodega Head and Malibu cases, the post-Oyster Creek atmosphere was one of buoyant optimism. All the participants in the regulatory process, industry and staff alike, seemed persuaded that the technology was basically in hand and that scaling up to commercial deployment would pose no unexpected or insurmountable difficulties. As Chauncey Starr, then president of Atomics International, imperiously summed up industry's position,

> Safety is a relative matter and I believe we have reached a point in the demonstrated safety of nuclear power to say nuclear power is safe, period. We can leave the questions concerning degree of containment and other technical safety discussions for those specific situations where detailed analysis is heard by professional people, who can make sophisticated judgment on the basis of understanding and knowledge of all the facts.[47]

Starr, a consummate believer in our technological prowess, went on to suggest that the regulatory bottleneck was of bureaucratic origin when he noted,

> If some acceptable risk level could be set as a target for our technical organizations, we could by both experiment and analysis determine the probabilities of a series of events and appropriately design protective devices where needed.[48]

Although the industry was already willing to make substantial economic commitments on the basis of those perceptions, the ACRS and the regulatory staff proved more reluctant. They wanted adequate experimental confirmation of the effectiveness of engineered safeguards before they relaxed their siting and design requirements sufficiently to permit urban siting or siting in the more active seismic areas.

To avoid the uncertainties and inequities imposed by subjective decisions, there was general agreement that the regulatory process would be greatly improved if decisions could be based on a set of quantitative siting and design standards. But because information on key scientific questions was poor and the pace of technological change was extremely rapid, no such set of standards could be drawn up.

The AEC had some means at its disposal for softening the impact of these factors. The Commission could have greatly strengthened the safety R&D capability and made it more responsive to regulatory needs. In fact, the Division of Reactor Development and Technology safety program was redirected to focus on maximum credible accidents, particularly those caused by loss of coolant. But the program included no work that could support the development of or substantiate industry-developed design and site requirements. Nor was the safety research and development program made responsive to regulatory needs, either through Commission intervention or through a reorganization that would give the regulatory staff some control over the program.

Second, the Commission was in a position to moderate the pace of commercial deployment and technological change through its rule-making and licensing powers. It might well have limited the number of plants licensed if the uncertainties seemed high. It also might have required a large degree of design standardization, so standards could have been developed more easily. But neither the Commission nor the regulatory staff appears to have seriously considered either of these options.

Instead the AEC let technological change and commercial use move with the dictates of the market. Then, to compensate for existing uncertainties, the regulatory staff seemed to adopt the most conservative requirements *consistent with the commercial viability of the nuclear power reactor*. The staff had no intention of seriously constraining its commercial use. But short of that, it would require what safety measures were available.

Notes

1. Joint Committee on Atomic Energy, *Improving the AEC Regulatory Process,* March 1961, p. 42.

2. Robert Perry and others, *Development and Commercialization of the Light Water Reactor, 1946-1976,* R-2180-NSF, June 1977.

3. Two other manufacturers, Combustion Engineering and Babcock & Wilcox, also sold PWRs, but neither contributed much to the technical evolution of the power reactor nor did either capture a significant share of the market during the 1960s.

4. Perry and others, *Development and Commercialization. . . ,* p. 91.

5. Ibid.

6. Ibid., p. 94. For a good discussion of the "turnkey phase" and early commercialization see Arturo Gandara, *Electric Utility Decisionmaking and the Nuclear Option,* R-2148-NSF, The Rand Corporation, June 1977.

7. Ibid., p. 51.

8. Ibid., p. 52.

9. Joint Committee on Atomic Energy, *Hearings on Licensing and Regulation of Nuclear Reactors,* 90th Congress, 1st Sess. (April 4, 5, 6, 20, and May 3, 1967), p. 659.

10. Ashley, *Nuclear Power Reactor Siting. . . ,* this figure included right-of-ways, two circuits, and capitalized losses, p. 100.

11. Ibid., p. 103.

12. "Development of Economic Nuclear Power," Memorandum, AEC 331/59, January 23, 1953.

13. *Nucleonics Week* 19:3 (March 1961).

14. Murphy, *The Nuclear Power Controversy,* p. 116.

15. Ashley, *Nuclear Power Reactor Siting. . . ,* 1965, p. 98.

16. Ibid., p. 117.

17. JCAE, *Radiation Safety. . . ,* June 1961.

18. JCAE, *Hearings on AEC Regulatory Problems,* April 1962.

19. Joint Committee on Atomic Energy, *Hearings on the Development, Growth, and State of the Atomic Energy Industry,* 1955.

20. Dawson, *Nuclear Power...,* p. 176. See Clifford Beck, "Engineering Out the Distance Factor," *Atomic Energy Law Journal,* 1963.

21. JCAE, *Hearings on AEC Regulatory Problems,* April 1962.

22. Joint Committee on Atomic Energy, *Hearings on Licensing and Regulation of Nuclear Reactors,* 90th Congress, 1st Sess. (April 4, 5, 6, 20, and May 3, 1967).

23. Ashley, *Nuclear Power Reactor Siting...,* 1965, p. 157.

24. Ibid., p. 55.

25. Green, "Safety Determinations in Nuclear Power Licensing: A Critical View," *Notre Dame Lawyer* 43:633 (June 1966), 642.

26. Ibid.

27. JCAE, *Radiation Safety...,* p. 275.

28. Ibid., p. 224.

29. JCAE, *Hearings on Regulatory Problems,* April 1962, p. 2.

30. JCAE, *Hearings on Licensing and Regulations,* 1967, p. 628.

31. Ibid., p. 64.

32. Ibid., p. 134.

33. JCAE, *Hearings on the Development, Growth...,* 1955, p. 203.

34. "AEC Steps Up Reactor Safety Experiments," *Nucleonics* 14:3 (March 1956), 45-47.

35. Joint Committee on Atomic Energy, *Hearings on AEC Authorizing Legislation, FY 1962,* 85th Congress, 1st Session, 1961.

36. Ibid.

37. Joint Committee on Atomic Energy, *Hearings on AEC Authorizing Legislation, FY 1963,* 87th Congress, 2nd Session, 1962, p. 98.

38. Ibid.

39. Ashley, *Nuclear Power Reactor Siting.* Loss of coolant would lead to overheating and fuel breakdown even if the reactor were immediately shut down.

40. See testimony of Joseph Lieberman, Assistant Director of Safety, DRDT, in Joint Committee on Atomic Energy, *Hearings on AEC Authorizing Legislation, FY 1965;* Part II, 88th Congress, 2nd Session, 1964, pp. 399 ff.

41. Ashley, *Nuclear Power Reactor Siting...,* p. 130.

42. See ACRS letter to the AEC, November 18, 1964.

43. Ashley, *Nuclear Power Reactor Siting...,* p. 134.   44. Ibid.

44. Ibid.

45. JCAE, *Hearings on Licensing and Regulation,* 1967, p. 215.

46. See Ashley, *Nuclear Power Reactor Siting...,* Session VII, "Public Acceptance of Nuclear Power," pp. 245-292.

47. Ibid., p. 6.

48. Ibid., p. 3.

6 Some Emerging Doubts

There is never any lack of articulation of the benefits of a technology. Every technology has powerful vested interests—private and frequently governmental and political—which can be relied upon to press the benefits [of the technology]. The problem is that the negative factors and the risks are never fully or even adequately articulated.[1]   —Harold Green, 1972

As nuclear power gained a commercial toehold, both the nuclear industry and the AEC seemed firmly convinced that the technology and its normal commercial development and use would present no unforeseen hazards. But few had really thought in terms of such rapid reactor scale-up or commercial deployment. As these trends became apparent, they suggested the possibility of a dramatic increase in public risk from nuclear reactors, and some members of the regulatory community began questioning the adequacy of certain design and safeguard characteristics.

Utilities had been contracting for an average of about one reactor per year since 1953. And most of those contracts depended on some AEC financial participation. While most participants in the nuclear business believed that General Electric's successful bid for the Oyster Creek contract heralded a new phase in the acceptability of nuclear power, no one anticipated the demand of the next several years. After one lean year during which utilities seemed to be taking stock, orders jumped from eight in 1965 to twenty-one in 1966 and then to twenty-seven in 1967 (see figure 6-1). The 1967 contracts represented 47 percent of all the generating capacity ordered that year.[2]

Not only were their numbers increasing at an unexpected rate, but so was reactor size. Between 1963 and 1967 the average reactor capacity jumped from 550 MWe to 850 MWe, the latter design representing almost a four-fold increase over the first commercial-scale reactors that came on line just a few years earlier (see figure 6-2).

As they grew, the new reactor designs also provided for better fuel burn-up characteristics and greater power density (see tables 6-1, 6-2, and 6-3). Designing larger cores with better fuel burn-up characteristics and trebling the fuel lifetime of a 1962 design meant there would be substantially greater quantities of fission products on hand in the reactor at any given time. For example, the decay heat (heat from the fission products when the reactor is "off") from the 1,065 MWe Brown's Ferry reactor contracted for in 1966 almost equaled Shippingport's full power output.[3] And increasing the power density of a reactor meant the fuel

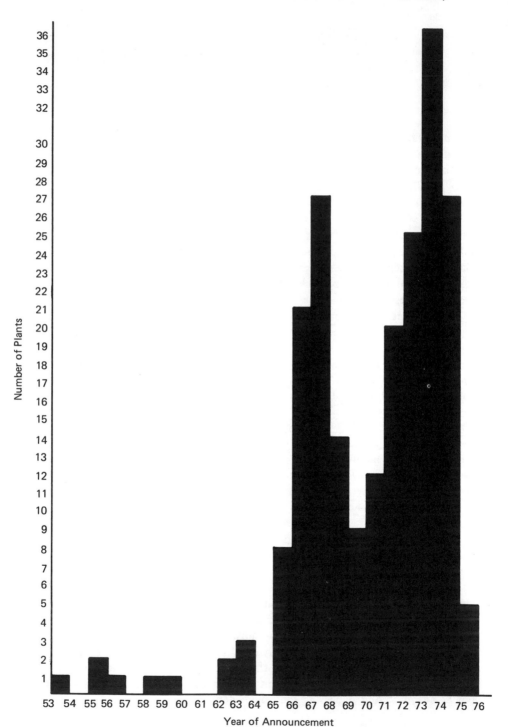

Figure 6-1. Nuclear Plant Order Announcements

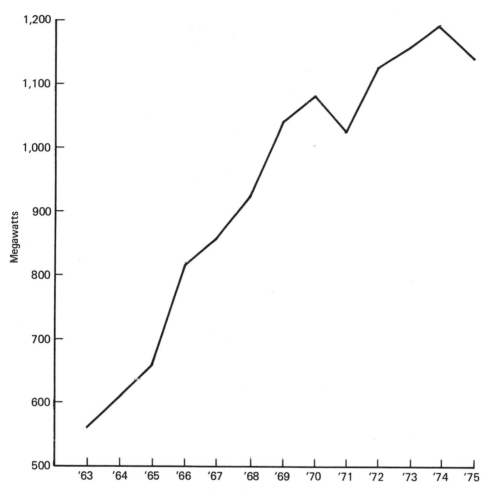

Figure 6-2. Average Capacity of Reactor Ordered by Year

would burn at a higher temperature. Therefore, in the event of a loss of coolant accident, the new designs required that *more* make-up cooling water be supplied more rapidly than did the older designs to obtain the same safety objective, preventing a core-melt.

Assuming no concomitant improvements in safety characteristics, the dominant development characteristics of the mid-1960s materially altered the risks imposed by nuclear power. If the risk is defined as the expected loss (that is, the probability of an accident multiplied by its consequences), then as the numbers of new reactors brought into use increased, so did the probability of a reactor accident. And as the reactor size and its inventory of fission products increased, so did the consequences of an accident.

Members of the regulatory community slowly became aware of these

Table 6-1
Westinghouse Plants—PWR

Plant	Start Year	Operating Year	Elec. Power, MWe[a]	Thermal Power, MWth[b]	Efficiency, %	Height, m[c]	Core Diameter, m[c]	Volume, m³[d]	Fuel Inventory, tonnes[e]	Power Density kw/l[f]	Power Density kw/kg[g]	Burn-up, MWd/te[h]	No. of Loops
Shippingport	1955	1958	185	600	30.8	2.3	1.9	6.5	20,700	92	28.9	25,000	4
Yankee (Rowe)	1957	1/61	600	1825	32.7	3.07	3.03	22.3	75,000	82	24.3	30,000	4
Haddam Neck	1/64	1/68	450	1350	33.4	3.07	2.82	19.0	57,000	71	23.6	30,000	3
San Onofre	5/64	2/68	470	1520	30.9	3.66	2.44	17.2	52,000	88	27.0	29,000	2
R.E. Ginna	2/64	5/70	730	2200	33.1	3.66	3.04	26.6	71,400	83	30.8	31,000	3
H.B. Robinson 2	4/67	1/71	820	2440	33.6	3.66	3.04	26.6	70,000	92	34.8	31,000	3
Surry 1	4/68	10/72	1050	3250	32.2	3.66	3.55	35.6	99,000	91	32.8	22,000	4
Zion 1	2/68	8/73	1130	3425	33.0	3.66	3.45	34.2	101,000	100	33.8	24,000	4
Trojan	10/69	9/75	1150	3425	33.6	3.66	3.45	34.2	95,000	100	36.0	25,000	4
Callaway City 1 (SNUPPS)	1976	1982											
South Texas	1976	1982	1250	3800	32.9	4.27	3.45	39.9	110,000	100	34.5	25,000	4

Source: R. Perry, *Development and Commercialization of the Light Water Reactor*, p. 89. Santa Monica: The Rand Corporation, R-2180-NSF, June 1977.

[a]MWe = Megawatts electric.
[b]MWth = Megawatts thermal.
[c]m = meters.
[d]m³ = cubic meters.
[e]tonnes = metric tons (1000 kg).
[f]kw/l = kilowatts/liter.
[g]kw/kg = kilwatts/kilogram.
[h]MWd/te = Megawatt days/metric ton.

Table 6-2
General Electric Plants—BWR

Plant	Start Year	Operating Year	Elec. Power, MWe[a]	Thermal Power, MWth[b]	Efficiency, %	Height, m[c]	Core Diameter, m[c]	Volume, m³[d]	Fuel Inventory, tonne[e]	Power Density kw/l[f]	Power Density kw/kg[g]	Burn-up, MWd/te[h]	No. of Loops
Dresden 1 (BR-1)	2/57	6/60	210	690	30.4	2.7	3.3	23.1	57,600	33	12.0	12,000	4
Oyster Creek (BR-2)	3/64	12/69	670	1930	34.7	3.66	4.07	47.6	125,000	40	15.5	15,000	5
Quad Cities (BR-3)	3/66	1/73	850	2510	33.8	3.66	4.82	67.0	155,000	39	16.2	19,000	2
Brown's Ferry 1 (BR-4)	2/67	7/74	1100	3290	33.4	3.66	4.82	67.0	169,000	49	19.5	19,000	2
La Salle 1 (BR-5)	2/71	1979	1120	3290	33.8	3.66	4.78	65.5	149,000	50	22.1	27,600	2
Perry 2 (BR-6)	10/74	1980	1250	3580	35.0	3.76	4.65	62.5	138,000	57	26.0	28,000	2

Source: R. Perry, *Development and Commercialization of the Light Water Reactor*, p. 93. Santa Monica: The Rand Corporation, R-2180-NSF, June 1977.

[a] MWe = Megawatts electric.
[b] MWth = Megawatts thermal.
[c] m = meters.
[d] m³ = cubic meters.
[e] tonnes = metric tons (1000 kg).
[f] kw/l = kilowatts/liter.
[g] kw/kg = kilowatts/kilogram.
[h] MWd/te = Megawatt days/metric ton.

Table 6-3
Combustion Engineering Plants—BWR

Plant	Start Year	Operating Year	Elec. Power, MWe[a]	Thermal Power, MWth[b]	Efficiency, %	Height, m[c]	Core Diameter, m[c]	Volume, m³[d]	Fuel Inventory, tonnes[e]	Power Density kw/l[f]	Power Density kw/kg[g]	Burn-up, MWd/te[h]	No. of Loops
Palisades	2/67	2/72	788	2200	35.5	3.48	3.35	30.7	84,000	72	26	26,000	2
Maine Yankee	11/68	1/73	860	2440	35.3	3.6	3.5	34.6	87,000	71	28	30,000	2
Calvert Cliffs	7/68	1975	845	2560	33.0	3.5	3.5	32.5	83,000	79	30		2
Arkansas 1-2	1/72	2/78	925	2760	33.5	3.81	3.45	35.5	87,000	78	32	33,000	2
San Onofre 2	6/74	6/80	1140	3390	33.7	3.81	3.45	35.5	89,000	95	38	34,500	2
Pilgrim 2	1974	1982	1150	3460	33.6	3.81	3.45	35.6	93,400	97	37	31,000	2
Palo Verde 1	1975	8/82	1300	3800	34.2	3.81	3.63	39.4	99,300	96	38	33,200	2

Source: R. Perry, *Development and Commercialization of the Light Water Reactor*, p. 95. Santa Monica: The Rand Corporation, R-2180-NSF, June 1977.

[a]MWE = Megawatts electric.
[b]MWth = Megawatts thermal.
[c]m   = meters.
[d]m³   = cubic meters.
[e]tonnes = metric tons (1000 kg).
[f]kw/l = kilowatts/liter.
[g]kw/kg = kilowatts/kilogram.
[h]MWd/te = Megawatt days/metric ton.

changes and their implications. In an intuitive effort to compensate for the escalating risks, certain groups within the community began to withdraw from the optimistic mainstream of opinion and to explore in greater depth new means for improving the safety characteristics of reactor designs.

Pressure Vessel Technology

The reactor's pressure vessel and primary system piping stood as the main barrier to the release of fission products into the outside environment. But by the mid-1960s the AEC still had adopted no design or maintenance standards to govern the manufacture and inspection of these key components. For more than fifty years the American Society of Mechanical Engineers had assumed responsibility for developing the standards used in manufacturing ordinary pressure vessels and piping. Most states had incorporated the society's standards into their own legally enforceable codes. But nuclear plants presented special problems. Quality control or assurance was more important in a nuclear than in a conventional plant. The materials underwent prolonged exposure to high levels of radiation. And because they were contaminated, reactor components were often hard if not impossible to inspect during the life of the plant.[4]

Moreover, as the reactor size grew, so did pressure vessel walls. In a 600 MWe reactor, the pressure vessel wall measured twelve inches thick. Industry had little experience with this type of heavy section steel technology and many questions regarding the behavior of the thick sections under stress and exposure to radiation remained unanswered by the mid-1960s. These uncertainties notwithstanding, the rupture of a pressure vessel was judged to be so improbable it was tagged an "incredible accident."

In 1965 the Advisory Committee on Reactor Safeguards met with representatives of seventeen different companies and organizations to review the safety of the pressure vessel in the later reactors. Following that meeting, the committee published a report questioning the adequacy of the pressure vessel design and recommending that more attention be paid "to methods and details of stress analysis, to the development and implementation of improved methods of inspection during fabrication and vessel service life, and to the improvement of means for evaluating the factors that may affect the nil ductility transition temperature and the propagation of flaws during vessel life." The committee also recommended that "means be developed to ameliorate the consequences of a major pressure vessel rupture" and went on to observe that "the orderly growth of the industry, with concomitant increase in number, size, power level and proximity of nuclear power reactors to large population centers will, in the future, make desirable, even prudent, incorporating in many reactors the design approaches whose development is recommended above."[5]

In response to the Advisory Committee's well-publicized concern, the Commission adopted pressure vessel standards in 1967. It also set up a long-term Heavy Section Steel Technology research program to explore some of the issues raised by the report. Several manufacturers also mounted research efforts. The research programs produced considerable new information resolving uncertainties about stress and flaw propagation. In addition, they led to the development and use of in-service vessel inspection techniques. With its 1965 report on pressure vessels, the ACRS launched a long and for the most part fruitless campaign for improved safety R&D. In this first instance, however, the Commission was responsive.

Adequacy of Emergency Core Cooling

All power reactors had cooling water make-up systems to replace normal operating losses (steam leaks, refueling losses, and so on). They also had extra cooling systems to remove the heat from fission product decay over the long term in case normal operation was suspended. But before 1966 no high pressure, high capacity system to provide make-up coolant in case of pipe break or similar accident existed. Instead, the regulatory staff concluded that the required complement of available safeguards (containment, systems to remove decay heat and reduce pressure buildup, and containment air clearing systems) could adequately contain the consequences of the possible fuel meltdown (see figure 6-3).

Early in 1966, the Division of Reactor Development and Technology directed the AEC's Nuclear Information Center to prepare a report on the "state of the art" of emergency core cooling systems.[6] Although that report was not completed for several years, and therefore could not have influenced the course of the early debate, the division's request suggests that the ability of available safeguards to contain a loss-of-coolant accident in the new larger reactors was coming into question.

By 1966, the licensing staff had also become increasingly concerned by the fact that it was having to apply assumptions and experimental data developed for the early 100 MWe reactors in technical assessment reports for the new 1,000 MWe plants.[7] The staff was acutely aware that the course and consequences of a loss-of-coolant accident were probably quite different from those postulated for the smaller reactors.

The issue came to a head over Consolidated Edison's application for Indian Point II, an 873 MWe pressurized water reactor with a very high power level. To control the postulated consequences of a loss-of-coolant accident, Indian Point II had to rely on an untried emergency core cooling system in addition to containment. Although the licensing staff finally recommended that a construction permit be granted, it did so with serious misgivings.[8]

At the same time the Advisory Committee on Reactor Safeguards became similarly apprehensive that a loss-of-coolant accident in the new generation of 800 MWe reactors could not be contained without adequate emergency core cooling. And the systems then in use were in no way demonstrated to be adequate. In June of 1966, the ACRS advised Commonwealth Edison that the

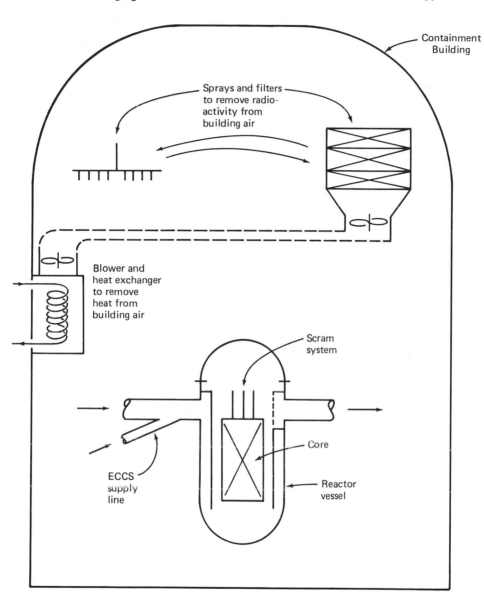

Source: Sidney Siegel, private communication.

Figure 6-3. Engineered Safeguards (of PWR)

committee could not approve its Dresden 3 design (809 MWe, BWR) because in the event of a loss-of-coolant accident, its unproven emergency core cooling system would be essential to containment.[9]

Both major reactor manufacturers briefly explored the possibility of actually coping with a core meltdown with a core retention crucible. But both rejected that approach as too costly, if indeed it was technically feasible. They proposed instead to improve their emergency core cooling designs. In August of 1966, the ACRS reported favorably on both the Dresden 3 and the Indian Point II applications, but recommended major improvements in the ECCS as well as in measures to reduce the probability of a loss-of-coolant accident (improving primary system quality, expanding in-service inspection, and improving leak detection capabilities).

However, the ACRS also concluded that for the future it would be prudent to develop a loss-of-coolant safeguard mechanism independent of the ECCS, particularly for reactors near population centers. In October of 1966, the committee sent a Reactor Safety Research letter to Chairman Seaborg recommending:

A vigorous research program should be initiated promptly on the potential modes of interaction between sizable masses of molten mixtures of fuel, clad, and other materials with water and steam, particularly with respect to steam explosions, hydrogen generation, and possible explosive atmospheres. Work should be directed toward understanding the mechanisms of heat transfer connected with such molten masses of material, the kinds of layers formed at cooled surfaces, the nature and consequences of any boiling of the fuel, and the manner and forms in which fission products escape from bulk molten fuel mixtures. Further, studies should be initiated by industry to develop nuclear reactor design concepts with additional inherent safety features or new safeguards to deal with low-probability accidents involving primary system rupture followed by a functional failure of the emergency core cooling system.[10]

The committee expected that an aggressive research and development program could produce a well-developed alternative or supplement to the ECCS within two years.

The commissioners, after reviewing the reactor manufacturers' position and after meeting with the ACRS, were not persuaded that an aggressive research program could produce a means for dealing with a molten reactor core. Nor were they persuaded that going that "next step" in limiting the consequences of a very low probability event would be a sensible use of their limited research and development funds. As an alternative to the ACRS's general recommendation, the Director of Regulation, Harold Price, appointed an independent task force headed by William Ergen from Oak Ridge National Labs and composed of twelve engineers, drawn equally from industry and from the AEC's research labs. The Commission charged the Ergen Committee (as it came to be known) with reviewing the loss-of-coolant accident and assessing the adequacy of available protective safeguard systems.

In October of 1967, the Ergen Committee presented the Commission with a 221-page report in which it recommended some possible improvements in the ECCS and concluded that the improved system combined with improvements in primary system quality adequately protected the public.[11] At the same time that it concluded the ECCS provided adequate protection, the committee recommended that the ECCS design be "systemized" and evaluated for reliability. The report also documented a number of uncertainties surrounding the functioning of the ECCS and the fuel rods during a loss-of-coolant accident and outlined a major experimental research effort that would close those information gaps.

Although the Ergen Committee was willing to support the ECCS approach, it clearly had reservations about the adequacy of the current design and about the regulatory staff's ability to evaluate a design. Moreover, the report clearly noted that even an improved ECCS might not always work, and while a partial core-melt would not necessarily breach the reactor's containment, a completely molten core would.

With somewhat ambiguous support from an independent group of experts, the Commission chose to live with the risk inherent in a large reactor with an improved ECCS. As David Okrent, then chairman of the ACRS, noted, ". . .It is fair to say that, at that time, the Task Force report provided those who were so inclined [with] a basis for rejecting any real effort on the question of core meltdown."

The Commission decision signaled a change in the standards of proof required by the AEC. The regulatory staff had required that PG&E be able to verify empirically that its Bodega Head design could withstand certain postulated seismic stresses. By accepting the Ergen Committee conclusions, the Commission set a precedent for accepting a higher level of uncertainty and for allowing analytical extrapolation to substitute for experimental data.

In its letter to the Commission on the Ergen Committee report, the ACRS repeated its recommendations for research and development work on the containment of a core meltdown.[12] In 1971, Dr. S.H. Bush and Dr. Joseph Hendrie, testifying before the JCAE on behalf of the ACRS, reiterated that position, noting the particular need for research results evaluating the effects of larger cores, greater power densities, and longer fuel lifetimes on loss-of-coolant accidents. The following year, after conducting some exploratory research of its own, the committee approached Milton Shaw, director of the Division of Reactor Development and Technology and the AEC's safety R&D program, again recommending research on core meltdown containment. The recent failure of the AEC's own semiscale tests of the ECCS notwithstanding (see below), Shaw advised the ACRS that he did not plan to sponsor any research on core-melt problems.[13] The nuclear industry consistently responded to the ACRS recommendation with similar disinterest.

The Safety Research Program

The safety issues emerging in the mid-1960s had two distinct components. First there was the question of what in fact was the risk? What were the physical principles underlying the event or sequence of events? What consequences would their unfolding have? How capable and reliable were the proposed safeguards? And second, what risk level was acceptable? Events forced the Commission to make subjective judgments on the latter question with considerably less than conclusive information on the former. When that is the case and the issue is at all controversial or there is significant disagreement regarding the judgment, prudence would dictate that relevant information be gathered even after the decision is made because, in all likelihood, the decision will be challenged.

A strong research and development capability would appear to have been invaluable whether it was used to identify hazards early on and determine the degree of risk they imposed or to verify decisions already made. But realities of politics and personalities seemed to preclude this obvious course. By the mid-1960s, industry had become an advocate of a strong, "practical," AEC-supported research program as a means of resolving defined engineering problems. Specifically industry wanted a program that included (1) research on the ECCS, (2) reliability studies for major safeguards and components, (3) definition of requirements for urban siting, and (4) probabilistic failure studies for safeguards and the overall reactor design. It viewed such a program as the obvious context for arriving at firm safety design criteria and standards that could then be applied in licensing. Conversely, the manufacturers and utilities adamantly opposed a loosely structured program devoted to exploring serious accidents and their consequences,[14] arguing that investing in the resolution of such low probability events contributed little or nothing to accident prevention or control and hence to the safety of the plant.

Industry found ready support from the new director of the Division of Reactor Development and Technology. In 1965 the Commission appointed Milton Shaw, a former aide to Admiral Hyman Rickover, to head the Division of Reactor Development and Technology (formerly the Division of Reactor Development). He was an able advocate, with the ear of both the Commission and the Joint Committee. And he came with several pronounced biases.

First, Shaw viewed light water reactor development as a job virtually completed and set his sights on the commercialization of a new reactor family, the liquid metal fast breeder reactor (LMFBR). He saw the LMFBR as the follow-on technology necessary to guarantee the fuel supply for all reactors when limited uranium deposits were depleted.

Second, Shaw was true to the Rickover philosophy that "an ounce of prevention is worth a pound of cure." He believed the basic power reactor design had proven itself to be safe, and the AEC's regulatory role lay in guaranteeing that the design was realized with high quality and reliable components and that the plant was run properly by competent personnel. Again like Rickover, Shaw believed in centralized management and a well-focused engineering approach to problem-solving.

Upon assuming responsibility for the division, Shaw redirected the programs under his control to support his priority objective, the development and commercialiation of the LMFBR, and its corollary, weaning industry from government-funded LWR research. He brought safety research under the personal scrutiny of his office. And during his first year, research projects were screened so vigorously that he failed to allocate his entire safety budget.[15] Robert Gillette, writing for *Science,* reported that many scientists working in the program believed that because of his commitment to the commercialization of the light water reactor and the LMFBR and because the full cooperation of industry was essential to the success of his LMFBR program, Shaw was reluctant to raise questions or conduct safety research that might slow commercial use of light water technology or irritate the nuclear industry.[16]

In 1966 Shaw staked out his first positive research priority: reequipping the Idaho Loss of Fluid Test Facility (LOFT) as a showcase for reliability or "quality assurance."[a] In the naval reactor program, Rickover could impose strict standards on his contractors and enforce them. In the civilian reactor program, the regulatory staff had, so far, been unable even to specify standards with any precision. By refitting the LOFT facility with components that met exacting and well-specified standards of performance and quality control, Shaw hoped to prove to the manufacturers, the utilities, and probably to the regulatory staff that such standards were definable and attainable.

Refitting the facility, of course, meant interrupting construction and substantially delaying its completion—not to mention the tests of loss of fluid and core meltdown effects. But Shaw, like industry, believed that an accident that would breach containment was extremely unlikely. He therefore may not have considered research into the adequacy of containment and the consequences of core meltdown useful and may have intended to redirect the loss of fluid test program.

Again, according to his critics, Shaw chose to make quality assurance a priority, not because it would contribute materially to plant safety but because it would improve plant availability.[b] Problems stemming from poor quality in reactors then coming on line were slowing construction and causing lower plant availability than had been expected. Both cost the utilities money and might well have dampened their enthusiasm for the new technology.[c] Shaw, however,

[a]As noted earlier, LOFT was originally designed to conduct a single core meltdown to test the adequacy of the containment structure and to verify industry's predictions of the time the meltdown took and of the fission products it would release.

[b]Availability is defined as that percent of the plant's total capacity that is "available" to generate electricity at any point in time. Availability declines rapidly when a reactor must be shut down or operated at reduced power levels because repairs are needed.

[c]See R. Gillette's series in *Science.* Quality control in nuclear plants was not necessarily *worse* than in fossil fuel plants, but it had to be better because the costs and risk of malfunction were so much greater.

persuasively argued that there was a "... close relationship which exists among safety, reliability, and economics, such that a plant which is capable of sustained operation according to prediction and plan generally will be economic and inherently safe."[17] His critics may have been extreme in their interpretation of his motives, but clearly Shaw had responsibilities as head of the Division of Reactor Development and Technology that competed with and sometimes conflicted with safety research objectives.

Shortly after Shaw undertook to transform the LOFT facility into a showcase for reactor quality assurance, the Commission changed the facility's test mission. Responding to ACRS and Ergen Committee recommendations the Commission redirected the project to address emergency core cooling questions rather than core meltdown questions. The Commission intended that the Power Burst Facility do complementary fuel research.

However, in spite of the urgency of the issue in many minds, neither LOFT nor the Power Burst Facility was pursued with particular vigor. Essential redesigning to accommodate the change in its mission combined with Shaw's continued efforts to make it a quality assurance showcase brought work on the Loss of Fluid Test facility to a near standstill for the next several years and increased its cost from $18 million to $35 million. The Power Burst Facility hardly fared better. It was completed four years late (1971) and also was 100 percent over budget.

While little progress was being made on the LOFT facility, it became increasingly evident to the regulatory community that the emergency core cooling issue had not been laid to rest, and that more research on loss of coolant accidents was needed. In February of 1970 the AEC published a "Water Reactor Safety Program Plan" (Wash 1146) identifying 139 unsettled safety questions and designating forty-four (many of which related to the ECCS) as "very urgent, key problem areas, the solution of which would clearly have great impact, either directly or indirectly, on a major critical aspect of reactor safety."[18]

Given the growing sensitivity of the problem, the Commission appointed a senior task force of four regulatory staff members headed by Stephen H. Hanauer to reevaluate the ECCS problem. From November 1970 until March 1971, Hanauer's group designed and supervised a series of semiscale tests, the results of which jarred industry and the regulatory community considerably. In the loss-of-coolant simulation tests, *all* the make-up cooling water was forced out of the "pressure vessel" along with the original coolant, leaving the "test reactor" with no cooling water at all. And while many researchers argued that the nonrepresentative size and design of the test device led to those results, the results were unexpected and therefore cast suspicion on other extrapolations from experimental ECCS data to the true situation.

In the face of the growing uncertainties, the task force drew up Interim Acceptance Criteria for emergency core cooling systems that specified considerably more conservative performance criteria and analytical techniques for determining performance than had been used in the past.[19] And the Commission, considering the semiscale test results sufficiently unsettling, put the Interim Acceptance Criteria into force without the customary thirty- to sixty-day comment period. It hoped its more conservative position would allay any concerns and the criteria could then be relaxed as good information became available.

During these later years, neither the Commission nor Shaw moved to speed up work on the Loss of Fluid Test facility or to augment significantly the light water reactor safety research budget. On the contrary, in February of 1971 Shaw, arguing budget shortages, canceled a key study being done at Oak Ridge National Labs on fuel rod behavior during a loss-of-coolant accident. Clearly, in spite of the attention being given them, he still did not consider low probability, high consequence events to be high priority research topics.

The overall safety R&D budget provides further evidence of the low priority accorded light water reactor safety research at that time. About 8.5 percent of the modest safety R&D budget went unspent between 1965 and 1968.[20] The budget increased somewhat over the following four years and was used (see table 6-4), but the share devoted to LMFBR safety research also increased from $5.3 million in FY 1969 to $20.4 million in FY 1972. During those same years, light

Table 6-4
Distribution of Nuclear Safety Budget, Fiscal Years 1964-1974
(dollars in millions)

FY Year	Light Water Reactor	Fast Breeder Reactor	Other[a]	Total
1964	$ 7.3	–	$9.0	$16.3
1965	10.6	–	8.4	19.0
1966	12.9	–	8.7	21.6
1967	20.1	–	8.5	28.6
1968	24.1	–	8.7	32.8
1969	21.0	$ 5.3	7.4	33.7
1970	22.1	7.7	7.3	37.1
1971	19.0	9.6	7.1	35.7
1972	22.4	16.8	2.4	41.6
1973[b]	29.8	20.4	2.4	52.6
1974[c]	30.3	30.3	3.6	66.0

Source: Private Communications with John Abbadesa, AEC Controller, June 28, 1973, as found in Frank G. Dawson, *Nuclear Power: Development and Management of a Technology*, p. 214. Seattle and London: University of Washington Press, 1976. Reprinted with permission.

[a]Detail for FBR not available for FY 1964 through FY 1968. Shown under "Other."

[b]Estimated expenditures as obtained from John Abbadesa. Actual budget was $55.4 million.

[c]Initial budget to Congress. The budget was increased to approximately $72.2 million. The proposed budget for FY 1975 was $93.0 million.

water reactor safety research funds remained about constant (see table 6-4). Considering the rapid inflation rate of this period, the light water reactor safety program clearly suffered real dollar reductions.

If the commercial reactor safety program lacked money, then, by implication, it lacked scope and support. What program there was mirrored the reactor development program's preoccupation with the reactor itself. There was little research into the hazards associated with various stages of the nuclear fuel cycle. Indeed, members of the Commission and the Joint Committee on Atomic Energy rarely even spoke of the fuel cycle and then only in the most reassuring terms.

The safety program's supporters were likewise few and far between. As AEC Chairman James Schlesinger later described it, Shaw, in spite of his unwillingness to be responsive to the expressed needs of the regulatory staff or to conduct certain kinds of research, was the program's strongest champion.[21] The regulatory staff, disagreeing often with Shaw's priorities and having no control over the program, took no position. On the other hand, the Bureau of the Budget, the Joint Committee, and even occasionally the Commission itself, argued that the technology was commercial and that industry should assume complete responsibility for demonstrating its safety.

The Advisory Committee on Reactor Safeguards viewed the AEC's flagging LWR safety research effort with alarm. In March and again in November of 1969, the committee called the Commission to task in very outspoken letters, expressing particular dismay over the shrinking R&D budget, the agency's failure to conduct several projects recommended by the committee, and the agency's sluggish responses to problem review requests. The letters then cited several research areas the committee thought warranted immediate attention, including partial or large-scale core-melting, fuel failure, seismic effects, and metropolitan siting criteria.[22]

By 1971, in testimony before the Joint Committee on Atomic Energy, the Advisory Committee on Reactor Safeguards described the situation as even more serious. There had been no progress on any of the recommended research topics, several programs had been canceled or delayed, and the R&D budget had suffered continuing erosion. Furthermore, the committee concluded that the industry clearly could not (and perhaps even should not) be held responsible for the research required to set regulatory standards.[23] It lacked adequate facilities and resources. And of perhaps greater consequence, the public would not have the measure of confidence in standards developed by industry that it would have in those developed by an independent body.

The Regulatory Community's Response
to Increasing Uncertainty

During the developmental and earliest commercialization phases, all members of the regulatory community believed their decisions were conservative in the extreme. They sited plants and approved designs that, even in the event of a major accident, could be expected to expose the public to negligible quantities of radiation. Before slowly relaxing the siting requirements, they demanded convincing empirical evidence that the compensatory safeguards performed adequately (to wit, the Malibu and Bodega Head decisions). While an enormous amount of acknowledged uncertainty still surrounded the physical principles of reactor technology and safety in those years, there was agreement in the scientific community that the "worst" consequences of a miscalculation could be contained. Hence, there was great certainty that the risk imposed by nuclear power was very low.

Once changes in reactor design and numbers substantially increased both the likelihood and the consequences of a serious accident and it became clear that existing containment designs could be breached, the regulatory groups were faced with several sets of questions. The first set centered on the availability and type of information needed to make defensible decisions. Did the information exist? Should it be objective or subjective, empirical or judgemental? If it did not exist, and the regulators were forced to make decisions on the basis of spotty or inconclusive evidence, should they acknowledge it? Or would that course unnecessarily jeopardize public confidence in the technology and the regulatory process? And finally, how much uncertainty should they allow before judging the risk "unacceptable" and halting free market development and use of the reactor?

When the Advisory Committee on Reactor Safeguards questioned the reliability of the pressure vessel, no one considered suspending licensings. However, the Commission and the nuclear industry acknowledged the need for more information and set up research programs to provide it reasonably promptly. But the Commission chose to deal quite differently with questions the Advisory Committee and the regulatory staff raised about the larger reactor's ability to contain a loss-of-coolant accident.

By almost refusing to support two construction permit applications, the Advisory Committee and the regulatory staff demanded action of some sort from the Commission. Because it seemed doubtful the existing safety technology could be markedly improved and because new information would be expensive and difficult to obtain, the Commission called for scientific experts to judge the acceptability of the risk rather than a research effort to define the limits of the risk. Moreover, the Commission made the Ergen Committee report available only on a limited basis, suggesting that the commissioners believed the uncertainties raised in the report would unduly alarm the public. Only when the uncertainties became more visible and demands for their resolution grew louder did the Commission take halting steps to get more objective information. It was a clear case of too little, too late.

Table 6-5 illustrates the patterns of change both in the level of uncertainty in emerging controversies and in the way the Commission chose to resolve the controversy.

Table 6-5
Patterns of Change in Controversies

	Controversies			
	Bodego Head	Malibu	Pressure Vessels	Emergency Core Cooling System
Who raised the issue	Regulatory Staff	Intervenors	ACRS	ACRS (Regulatory Staff)
Perceived degree of uncertainty	Low	Low	Low-moderate	Moderate
Source of information used to resolve uncertainty.	Industry	Industry	AEC, Industrial R&D	Expert opinion
Character of information (subjective, objective) used to resolve uncertainty.	Objective	Objective	Objective	Subjective
Was uncertainty acknowledged	Yes	NA	Yes	No
Who made decision	NA	Commission	Commission	Commission

The regulatory staff apparently did give brief consideration to suspending licensing following publication of the Ergen Committee report.[24] But such qualms were transitory. The general decision-making principle of the early 1960s seems to have held: be as conservative as possible short of disrupting the commercialization process. As the uncertainties grew, the Commission chose to live with them.

A second set of questions centered on the source of the information. Should the AEC assume the responsibility for and cost of conducting a safety R&D program that was fully responsive to regulatory needs? Should it rely instead, or to some degree, on industry to produce the experimental evidence? Or should it settle for the subjective judgment of experts?

Clearly the Commission asked very little of its own safety R&D program during those years. It supported no appreciable budget increases for the program. It allowed the program to operate independently of the regulatory community. Although a steering committee made up of representatives from the regulatory staff, industry, and the Division of Reactor Development and Technology had been set up in response to the Mitchell panel's recommendation that the safety R&D program be more directly answerable to regulatory needs, the steering committee could only suggest, not require. Director Milton Shaw decided priorities and projects. And his choices seemed increasingly unresponsive to the pressing regulatory needs as defined by the ACRS and the regulatory staff.

The Commission seemed quite uninterested in developing a strong information-gathering capability of its own either to anticipate problems or to independently verify information from industry. Nor did it seem aware of the suspicion with which outsiders might view judgments based on the industry's evidence.

A third set of questions was implicitly raised by the conflicting approaches to safety pursued by the Advisory Committee on Reactor Safeguards and Milton Shaw. If risk is made up of a consequence component and a probability component, which had priority when allocating research resources? Should the Commission, at some point, admit to and accept the existence of low probability-high consequence accidents, and concentrate its safety resources on the higher probability, lower consequence events? If such a course promised a higher pay-off in measurable benefits, would it be politically feasible? The commissioners never explicitly acknowledged their dilemma, but in the case of the emergency core cooling issue, they seemed to conclude that the risks posed by a low probability-high consequence accident were acceptable. And by allowing Shaw to pursue his own R&D program in conflict with the recommendations of the Advisory Committee and the Ergen Committee, they lent tacit support to Shaw's position.

Notes

1. Raphael Kasper, ed., *Technology Assessment: Understanding the Social Consequences of Technological Applications* (New York: Praeger Publishers, 1972), p. 51.

2. Joint Committee on Atomic Energy, *Hearings on Nuclear Power Plant Siting and Licensing,* 93rd Congress, 2nd Session, 1974.

3. JCAE, *Hearings on Licensing and Regulation...* , 1967, p. 95.

4. AEC, *The Safety of Nuclear Power Reactors and Related Facilities,* Wash-1250, July 1973, pp. 7-9.

5. ACRS Report, November 24, 1965, as found in JCAE, *Hearings on Licensing and Regulation,* 1967, pp. 118-119.

6. See C.G. Lawson, *ECCS for Light Water Reactors,* ORNC-NSIC-24, October 1968.

7. Gillette, "Nuclear Safety," I-IV, *Science* 177 (September 1, 8, 15, and 22, 1972), 972.

8. Ibid.

9. David Okrent, "Some Thoughts on Reactor Safety," preprint, p. 2.

10. ACRS Report of Safety Research Programs (Okrent to Seaborg, October 12, 1966).

11. See W.K. Ergen and others, *Emergency Core Cooling, Report of Advisory Task Force on Power Reactor Cooling to the AEC.* Unnumbered Report, October 1967.

12. ACRS Report on Advisory Task Force on Power Reactor Emergency Cooling (Zabel to Seaborg, February 26, 1968).

13. Letter from M. Shaw, DRDT to R.F. Fraley, dated February 3, 1972, as noted in Okrent, "Some Thoughts on Reactor Safety."

14. JCAE, *Hearings on Licensing and Regulation...* , 1967. For a strong statement of their position, see letter from Atomic Industrial Forum to Shaw outlining suggestions for the agency's safety program, pp. 435-438.

15. R. Gillette, "Nuclear Safety," p. 77.

16. Ibid.

17. Joint Committee on Atomic Energy, *Atomic Energy Authorization Hearings FY 1971.*

18. *Water Reactor Safety Program Plan,* USAEC, WASH-1146, February 1970.

19. See *The ECCS Rule Making Hearings,* Docket R-M-50-1, 1971-1973.

20. For a full account, see Gillette, "Nuclear Safety...."

21. Joint Committee on Atomic Energy, *Hearings on AEC Authorizing Legislation FY 1974,* 93rd Congress, 1st Session, 1973, p. 33.

22. For the full text of the Advisory Committee on Reactor Safeguards letters to Robert Hollingworth, General Manager of the AEC, see JCAE, *Hearings on Licensing Procedure...* , 1971, pp. 1111-1121.

23. See Testimony, Dr. S.H. Bush, Advisory Committee on Reactor Safeguards, ibid., pp. 114-115.

24. S.M. Stoller Corporation, *Central Station Nuclear Power,* March 1976, unpublished.

7 Years of Encounter

This man Comey, representing that group, appeared before this committee, and among his qualifications that he listed, as I remember, was that he was a member of the Episcopal Church, and a Republican, and two or three other things, none of which had anything to do with expertise in the field of nuclear energy. As far as I am concerned, and in the view of most of the members, he was completely discredited as being an expert in the field.[1]–Representative, Chet Holifield, 1972

No elite group of experts, no matter how broadly constituted, has the ability to make an objective and valid determination with respect to what benefits people want and what risks people are willing to assume in order to have the benefits.[2]
–Harold P. Green, 1975

By 1968, technological evolution and rapid commercial adoption of nuclear power coupled with the serious safety issues voiced by the ACRS and the Ergen Committee were sufficient to stress the regulatory system severely. And that burden was substantially increased by changes in the outside environment. Commercial nuclear power came of age in a moment of considerable national ferment. Confrontations over the means of achieving social change and later over the U.S. role in Vietnam reflected an increasing unwillingness of some substantial segment of the public to allow "authority" or "expert opinion" to determine the course of U.S. policy unchallenged.

At the same time, the environment emerged as a dominant national concern. And like their predecessors in the antiwar movement, the environmentalists became skilled activists, confronting government and industry on a variety of issues, through a multitude of legal and legislative channels and at all levels of government. As they worked, environmental objectives gained currency and environmental groups gained political experience and sophistication. By the early 1970s, they had an impressive string of victories to their credit, including the National Environmental Protection Act (1969) and the Water Quality Improvement Act (1970).

For several reasons, nuclear power offered a particularly attractive target to environmentalists. First and foremost, it did indeed raise some genuine and serious environmental questions, at least one of which (radiation exposure levels) was assuredly dramatic. Second, because the general public was increasingly interested in environmental issues, more and more books and articles documenting (or purporting to document) the hazards and controversies surrounding nuclear power appeared in the popular literature. In their turn, these writings generated more public interest and concern for this particular environmental issue. Third, the technology was new and its hazards were not yet familiar. Its newness also meant it was less well entrenched and politically defended than an older industry. It, therefore, presented a more vulnerable target. Last, and most ironically, the rather unusual openness of the AEC's licensing process attracted opponents of nuclear power. The ordinary citizen was given formal access and considerable procedural leverage at an apparent decision point, the licensing hearing.[a]

Because the environmentalists were the latecomers and an "external force" beyond anyone's control, industry and the AEC tended to credit them with considerably more influence over licensing outcomes than can be documented. From 1962 through 1966, the AEC received twenty-six applications for construction permits, of which three (12 percent) were contested.[b] From 1967 through 1971, utilities submitted another seventy-four construction permit applications, twenty-four (32 percent) of which were contested on grounds of inadequate safety or adverse environmental effects.[c] Another five operating license applications were contested during these later years. And of the 100 applications filed between 1962 and 1971, fifteen eventually reached the courts. Clearly, although interventions were increasing in number, they were by no means the rule.

In many instances, especially during the early years of environmental opposition, government agencies (federal, state, and local) were the predominant and most effective intervenors.[d] They had the experience and resources to carry the battle through the hearing process and to the courts if need be. Through the late 1960s, individuals and small local groups relying on spare time and free local talent represented private environmental interests. Over those years they gained considerable expertise both in the basics of nuclear technology and in the manipulation of bureaucratic regulations.

As the environmental movement matured and money and talent became available for public interest activities, environmental intervention took on a somewhat new character. Some permanent local and regional groups were organized, occasionally staffed by a paid executive.[e] National environmental

---

[a]In the bitterly contested Monticello (Minn.) case (see following), a local high-school student availed himself of the broad access offered, got standing as an intervenor, and fought licensing of that plant to the very end.

[b]In addition to the Bodega Head and Malibu interventions on environmental and seismic grounds, the application for Turkey Point 3 (Dade Co., Fla.) was unsuccessfully opposed on the grounds that the plant was not designed to withstand sabotage or acts of war.

[c]The number does not include the four reactors contested on antitrust grounds.

[d]Instances where application was made during this period and government agencies played a major adversary role include: Vermont Yankee, Monticello, Zion 1 and 2, Calvert Cliffs 1 and 2, Indian Point II, Quad Cities 1 and 2, and Turkey Point 3 and 4.

[e]Major examples were David Comey's Businessmen for the Public Interest (BPI), the Lloyd Harbor Study Group, and the New England Coalition.

organizations and consumer groups began joining local groups in opposing specific plants, particularly where they saw an opportunity for a precedent-setting victory. Increasingly they hired seasoned legal talent to represent their interests both in hearings and in court, if necessary. By the early 1970s, technical expertise also became more readily available, as scientists with reservations about the safety of nuclear power organized to help the environmental groups and to intervene on their own initiative.[f] Notwithstanding this limited use of professional talent, environmentalist participation in the regulatory process remained predominantly a grass-roots, voluntary phenomenon, organized locally on an application-by-application basis.

Environmental intervention had several goals, some of which changed as the groups became more familiar with the possibilities and limitations of the system. At first intervenors pursued the simple objective of persuading the AEC-appointed Atomic Safety and Licensing Board to require a plant modification or site change, but to little avail. In fact, Southern California Edison's Malibu plant was the last project to be successfully opposed in a hearing. Although environmentalists had little chance of winning, they learned to delay plant construction and then to use the threat of delay to bargain independently with the utilities for design changes.[g] Hearings also provided the environmentalists with a forum for ecological education and publicity. And finally, although the debate centered on the safety or environmental effects of a particular plant, as the rules required, many of the groups participating saw intervention as the best and perhaps only means of blocking the diffusion of all nuclear technology.

For the first time in the regulatory history of commercial nuclear power, what constituted an "acceptable" risk became a major subject of *public* controversy, highlighting some different aspects of the regulatory process. The post-1968 years, therefore, provide a good opportunity to examine if and how the regulatory process resolves differences over the accuracy of information and differences over what constitutes an "acceptable" level of risk. How fairly the conflicts appear to have been resolved is also an important issue. Moreover, for any resolution to be reasonably permanent, it must, in some sense, be acceptable to all major interest group participants in the controversy. For the regulatory agency to enjoy sufficient public trust and respect to remain politically viable, it cannot ride roughshod over very sizable minorities even in the interest of the majority.

The Issues

Although there had been instances of public intervention in licensing proceedings prior to 1967 (the Fermi, Bodega Head, and Malibu plants), they were

---

[f]The Union of Concerned Scientists and the Scientists' Institute for Public Information were the most prominent of these groups.

[g]Though sound figures are hard to find, one source estimates the utilities lost about $1 million per month of delay in operation of the large plants. William T. Keating, "Politics, Energy and the Environment," *American Behavioral Scientist* 19:1 (September 1975), 60.

isolated and reflected no real trend. By the late 1960s, however, the situation changed dramatically. Not only were interventions increasingly common, but opposition to nuclear plants centered on a handful of specific issues.

Thermal Pollution

The first issue to emerge during this period was thermal pollution. Nuclear generating plants were considerably less efficient than fossil plants in that only about 30 percent (compared with 40 percent for fossil plants) of the heat produced could be translated into electricity. The remaining 70 percent had to be dissipated as waste heat through cooling water. Although the technology for closed cycle cooling towers existed, towers added $5 to $6 million to the cost of a plant.[3] Rather than pay the extra cost, utilities invariably chose to locate on waterways where the more abundant supply of cooling water could be used simply to carry the waste heat "away."

The mechanisms and environmental effects of thermal pollution were easily understood by the layman. The warmer temperatures of the power plant effluent cooling water raised the temperature of waterways into which it was dumped. The temperature increases, in turn, threatened to alter the ecology of the waterway, killing off some fish species, providing a more hospitable environment for others, and supporting faster-growing vegetation. Broad segments of local communities, including fishermen, water sports enthusiasts, and related tourist industries felt threatened by the potential effects of thermal discharges on their waterways. Hence thermal pollution offered a logical point around which opposition to nuclear plants could coalesce.

Intervenors raised the issue of thermal pollution in a number of proceedings. They fought their first and major contest over Vermont Yankee. Vermont Yankee Nuclear Power Corporation applied for a construction permit for its 514 MWe boiling water reactor in December of 1966. Vermont Yankee, located on the Connecticut River, was designed for "once through cooling," and it was expected the outfall water would raise the nearby river temperature about four degrees. Vermont's Department of Fish and Game demanded the water be returned at river temperature to avoid altering the ecology of the area, and in the 1967 hearings Vermont, New Hampshire, and Massachusetts opposed the issuance of any construction permit without that constraint. The AEC responded adamantly that its purview was limited to radiological characteristics and that it had no jurisdiction over the thermal qualities of reactor effluent. In 1969 the Boston Court of Appeals upheld the AEC's contention, and later the Supreme Court denied New Hampshire's petition for review.

Intervenors were undaunted by the regulatory staff's early reluctance to consider thermal effects. Between 1966 and 1971 three state governments (Illinois, Minnesota, and Maryland), the United Auto Workers, and several environmental groups intervened in a total of fourteen licensing hearings (construction permit and operating license) on grounds of thermal pollution. But in all instances the Atomic Safety and Licensing Board refused to hear arguments on the subject.[h] However, the intervenors did not go home empty handed. Well before legislation forced the AEC to accept jurisdiction over environmental effects (see below), the threat of costly hearing delays prompted the utilities to turn reluctantly but voluntarily to cooling towers or ponds. The trend is illustrated in figure 7-1.

Perhaps the best example of the self-consciousness with which the environmentalists exploited their strength is seen in their battle with Consumer Power Company over the operating license for that utility's 700 MWe Palisades plant on the eastern shore of Lake Michigan. After forty days of hearings and threats of prolonged court action, Consumer Power opened private negotiations with the environmentalists. After another forty days of negotiations, the utility concluded an eighty-two-page contractual agreement that bound it, among other things, to install cooling towers and "zero release" equipment to control radioactive effluents. In return, the environmentalists withdrew their objections to the license.

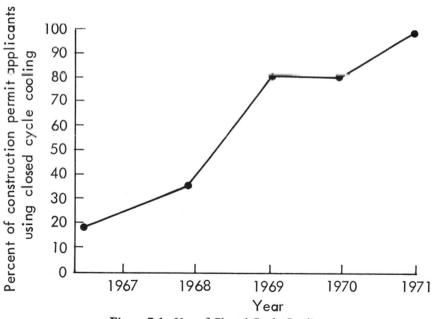

Figure 7-1. Use of Closed Cycle Cooling

---

[h]Other major contests include Turkey Point 3 (Florida), Palisades (Michigan), Monticello (Minnesota), Shoreham (Long Island), Indian Point II (New York), Zion (Illinois), and Calvert Cliffs (Maryland).

The Joint Committee on Atomic Energy, more sensitive than the AEC to the changing tides of public opinion, introduced legislation during the 1968 session to give the AEC control over thermal effluent standards. But the AEC, unwilling to take on this new responsibility, opposed the measure and it was dropped.

Then in 1969 Congress wrested the matter from the hands of the nuclear power establishment. The National Environmental Policy Act was passed that year and became effective in January of 1970. Section 102(2)c of the act required all federal agencies with licensing authority to consider the environmental consequences when licensing a facility or activity. For each plant the AEC had to produce an Environmental Impact Statement specifying (1) the expected environmental effects of the plant, (2) its adverse effects, (3) the alternative means and effects of producing the power, and (4) any irreversible commitment of resources if the license were granted.

The AEC, still eager to avoid responsibility for environmental conflicts, asserted that where federal, state, or regional environmental standards already applied, it would not assess the environmental effect of the plant. It would only determine whether the plant met the existing standards. Otherwise, the AEC agreed to draft full Environmental Impact Statements in compliance with NEPA, but only for plants applying for licenses after February 1971, arguing the year's intervening time was necessary to comply with the act. The agency further stated that consideration would be given to environmental issues at hearings only when they were contended.

The Calvert Cliffs Coordinating Committee, an environmental coalition of several smaller groups opposing construction of Baltimore Gas and Electric Company's two 845 MWe units at Calvert Cliffs, Maryland, appealed the AEC's limited interpretation of the responsibility placed upon it by NEPA. In July of 1971 the court of appeals for the District of Columbia handed down a landmark decision ruling in favor of the Coordinating Committee. The court observed that "the Commission's crabbed interpretation of NEPA makes a mockery of the Act."[4] The decision required the AEC to (1) honor NEPA provisions in all licensing cases begun after January 1, 1970, (2) conduct independent assessments of water quality and other environmental factors at all important decisionmaking stages, and (3) independently review all NEPA topics in uncontested as well as contested hearings. Not only would the AEC have to assess the environmental effects of all plants applying for licenses, but it would have to do so retroactively for the more than thirty reactors that had applied for licenses during the previous two years.

While the AEC stood firm and the Calvert Cliffs Coordinating Committee sought redress in the courts, Congress and the president did not ignore the growing public support for the environmentalists' position. In 1971, Congress passed the Water Quality Improvement Act, giving AEC unequivocal responsibility for thermal effluent characteristics of the plants it licensed. That same year President Nixon ordered the Army Corps of Engineers to investigate the possibility of requiring discharge permits in accordance with the Navigable Waters Refuse Act of 1899 and threatened to use this obscure provision against nuclear plants.

In 1972 Congress amended the Federal Water Pollution Control Act over a presidential veto. The new provisions made the states responsible for setting and enforcing effluent standards that were acceptable to the Environmental Protection Agency (EPA). If the states failed to meet the EPA's standards, then the EPA was empowered to assume responsibility itself.[5] Since the Environmental Protection Agency was thought to be the government agency most sensitive to environmental problems, Congress further supported the position of the intervenors in their fight with the AEC by centralizing effluent control in that agency.

Radioactive Effluents

The controversy over thermal effluents represented a clear difference of opinion between the intervenors and a majority in Congress, on the one hand, and the utilities and the AEC, on the other, over the costs and benefits of closed cycle cooling. The effects of increasing water temperatures were generally understood and the technology to eliminate thermal effluents was available. It remained to be decided whether those effects were sufficiently "bad," on balance, to warrant imposing the extra costs of closed-cycle cooling on the utility's rate-payers.

The controversy over what levels of radioactive effluents to permit under normal plant operating conditions represented a much more complex dispute. Not only was the acceptability level at issue, but there was also great uncertainty and debate over what risk low-level exposure actually imposed.

Under normal operating conditions nuclear plants inevitably lose small quantities of radioactive material to the outside environment. These normal, low-level releases pose two problems. First, direct exposure to them may constitute a hazard. Second, they may spread through the environment and concentrate in the food chain, thereby increasing the rate of exposure for people consuming the contaminated foods. Through a variety of technical devices, none of them terribly costly, low-level releases could be controlled.[i] To determine what control strategies should be required, the AEC needed to determine what risk burdens it believed low-level emissions imposed. To assess the risk burden, the AEC had to determine what exposure levels were actually harmful to man and then what emission levels would result in what exposure levels.

---

[i]Such devices include low-pressure containment, scrubbers, and gas hold-up systems. They were estimated to cost less than 1 percent of the capital cost of the plant, plus operating expenses of another $1 to $2 million per year. See E.C. Tsivoglou, "Nuclear Power: The Social Conflict," *Environmental Science and Technology* 5:5 (May 1971), 410.

In 1929 the Advisory Committee on X-Ray and Radium Protection, a self-appointed group of radiologists headed by Dr. Lauristan Taylor, was set up as an adjunct to the Bureau of Standards. Dr. Taylor also built and ran the bureau's laboratory, the only one in the country devoted to radiation exposure research.[6] The Advisory Committee represented the United States on the International Commission on Radiological Protection and over the next decades adopted the standards agreed on by the International Commission. In time, the Advisory Committee, later named the National Committee on Radiation Protection and Measurement, became a quasi-governmental body. Although Congress did not give the group statutory authority, it continued to represent the United States at the international level, and until 1960 it remained the only source of exposure standards used in this country.

From its beginning, the National Committee on Radiation Protection accepted the International Commission's assumption that there was a "threshold" or level of exposure below which man experienced no ill effects. It was thought that if exposure was held below some level, damaged cells could repair themselves. The crude kinds of laboratory experiments possible in the 1930s and 1940s appeared to confirm this assumption. As the years passed and measurement and experimental techniques improved, levels at which effects could be detected became lower and recommended exposure limits were consequently reduced. But the underlying assumption—that there was an exposure threshold below which man would suffer *no* ill effects—persisted.

Until 1955, the National Committee on Radiation Protection only issued standards governing occupational exposure. That year, as a result of growing public concern over radioactive fallout from nuclear tests, the committee adopted exposure standards for the general population—an arbitrary 10 percent of the permissible occupational exposure level. And in response to questions being raised in the Fermi plant hearings, the committee set nuclear power plant environs standards at 30 percent of the permissible occupational exposure level in April of 1958. It made sense that general population and plant surroundings levels should be lower than the occupational standard, because more people would be at risk and their exposure would be involuntary. At the same time, the nuclear industry could meet these lower levels with the technology then in use. Those were the only criteria to be applied, and the new standards were essentially pulled out of thin air.

But during these same years, there were two major accidents involving radioactive material. In March of 1954, ash from the Bikini nuclear tests rained down on the fishing crew of the *Lucky Dragon,* exposing and ultimately killing many of her crew. Then in October of 1957, the British test reactor at Windscale suffered a uranium fire and containment failure that spread radioactive material over the surrounding countryside where it rapidly invaded the food supply. These well-publicized events coupled with growing levels of radioactive fallout from nuclear testing prompted increasing public concern in the United States about the adequacy of the National Committee on Radiation Protection's standards.[j]

At the same time, the scientific community began to question the assumptions and the research that supported the National Committee exposure standards. In 1956 a National Academy of Science report suggested there might not be a threshold below which exposure had no effect. Instead, cumulative exposure might be the key consideration. Two years later the United Nations Scientific Committee on the Effects of Atomic Radiation supported the National Academy report. Later a member of the International Commission on Radiological Protection noted that the research traditionally used to establish tolerance levels—almost always single-dose experiments on animals—was bound to support the quite possibly erroneous threshold hypothesis.[7]

As the National Committee standards, also formally adopted by the AEC, became suspect, the surgeon general, Dr. Leroy E. Burney, formed the National Advisory Committee on Radiation to evaluate current handling of radiation health problems. When the Advisory Committee criticized existing standards and recommended that a central regulating authority be created, Burney moved to get authority over radiation standards transferred from the AEC to the Department of Public Health. But, in spite of strong backing from Senator Lister Hill, he was unsuccessful.

In two later moves the National Committee on Radiation Protection and the AEC further jeopardized their credibility as impartial arbiters of exposure standards. In 1959, when the International Commission tightened its standards for ingested radioactive material, the National Committee relaxed the standards that applied in the United States. The National Committee said it was simply figuring the dosages on "a new basis."[8] But nuclear testing had raised the level of radioactivity in foodstuffs and water in many states in excess of the new International Commission standards. And since the new American standards accommodated all the observed increases, there was some reason to suspect that the National Committee's new standards were founded on expediency rather than scientific merit.

That same year, Senator Ralph Yarborough of Texas attacked the AEC's newly published low-level waste disposal plans. The AEC had selected sites along both coasts and the Gulf of Mexico, where it intended to dump accumulating low-level wastes. Some sites were very near urban areas and in currents or in waters less than 100 feet deep. As the first challenge from an articulate opposition disclosed, neither Lauriston Taylor nor the AEC had a clear appreciation of the probable consequences of such disposal practice.[9]

The opposition was fierce enough to force the AEC to abandon most of the proposed disposal sites. However, the Commission adamantly refused to give up its Gulf sites, in spite of appeals from numerous Gulf states and their congres-

---

[j]As one measure of public concern, in 1959 six major sets of congressional hearings were held on the AEC, four of which dealt with aspects of exposure to radiation.

sional representatives. Finally, Senator Lyndon Johnson persuaded the Mexican ambassador that the danger applied equally to the Gulf coast of Mexico, and the ambassador filed a formal diplomatic protest. At the request of the Department of State, the AEC finally abandoned its Gulf water disposal sites.[10]

Given the public's growing concern over the increasing burden of radioactive material in the environment, President Eisenhower was anxious to end the struggle between the Public Health Service and the AEC for control of exposure standards as quickly and amicably as possible. To this end he created a Federal Radiation Council in August of 1959.[11] The council members included the Secretaries of Health, Education, and Welfare, Defense, Agriculture, Interior, Commerce, Labor, and the Chairman of the AEC. The council was to advise the president on standards policies and the organization of executive efforts to control ionizing radiation. It was to use its own advisors and the findings of the National Committee and the International Commission to develop exposure guidelines applicable to all federal agencies, including the AEC. The president hoped that this new high-level council could conclude any debate over standards in the official world and, by virtue of its rank, lend more credibility to the standards selected.

But the Federal Radiation Council was equipped to do little more than smooth troubled political waters. It had no research capability and its members had no expertise in radiological matters. Within a few months of its creation, the Federal Radiation Council decided to permit agencies to develop their own standards if they preferred. Although it did issue several reports containing new radiation protection guides, the council continued to rely on the National Committee for guidance, and the guidelines were not seriously applied or enforced.

With the signing of the Nuclear Test Ban Treaty in 1962, public concern over ambient radiation levels subsided and the AEC continued to depend on the National Committee on Radiation Protection's exposure standards. Through the early 1960s dose measurement and descriptions became more sophisticated. Exposure levels for various vital organs were set, independent of the whole body allowances, and the recommended levels were reduced somewhat. But the agency's own biological research program remained relatively modest, and no epidemiological studies were undertaken.

There was, however, a steady shift away from the "threshold hypothesis." Instead, the scientific community argued it was at least prudent and probably accurate to assume that *any* exposure level posed a possible health hazard. Because the mechanisms of radiation damage were still not understood and because the effects of low-level exposure were too small to be experimentally detected, scientists began calculating the risks from exposure to low levels of radiation by extrapolating from the effects of higher doses where the effects were measurable. They believed this to be the most conservative estimating technique.

The "linear effect hypothesis" posed a serious potential difficulty for the AEC. Once it was accepted that *any* level of exposure posed a hazard, then complete safety could not be guaranteed. Instead, the AEC would have to decide what risks people would be willing to accept in return for what benefits and to defend its judgments. And instead of determining the exposure "threshold" and forcing industry to control radioactive effects accordingly, the Commission would have to insure that industry did the necessary R&D and compromised its cost objectives sufficiently to keep emission "as low as practicable." Moreover, the Commission might have to persuade a skeptical public that the effluents were indeed being held as low as was reasonable.

As the scientific thinking changed, the Commission did admonish the nuclear industry to keep operating emissions "as low as practicable." At the same time, however, the Commission adopted exposure standards (10CFR20) that permitted radiation releases well in excess of what the best technology of the period could control. Industry and the regulatory staff argued that these higher levels presented no apparent health risk and allowed needed flexibility, since complex systems always vary in performance. For all practical purposes, the higher levels were to permit temporary violations of the "as low as practicable" standard.

Perhaps of greater consequence, neither the Commission nor the regulatory staff called attention to the shifting assumptions about the effects of low-level exposure. They continued to argue that the standards were "safe": that is, "risk-free." In fact, they often observed that the standards were sufficiently conservative to offer "a many-fold margin of safety."[12] Such an assertion implied that the AEC continued to operate on the assumption that there was an exposure threshold below which no effects occurred, and the standards held exposure well below that threshold. Clearly the AEC was not yet prepared to discuss risks versus benefits.

In the late 1960s, Dr. Ernest J. Sternglass, a University of Pittsburgh physicist, began publishing the results of several infant mortality studies he had conducted in the neighborhood of operating reactors. He had found what he argued was a causal correlation between the operation of nuclear reactors and an increase in infant mortality. He maintained his data conclusively demonstrated that the consequence of exposure increased as a positive function of dosage. That is, there was no threshold or level below which exposure was "safe." The "Sternglass Correlation" got some limited attention in the popular literature, and the AEC, fearing its exposure standards might be challenged, moved to develop some experimental data to support its own position.

John Tamplin, head of a major biomedical program at the AEC's Livermore Radiation Laboratory, and his colleague Arthur Gofman, were asked to evaluate the Sternglass conclusions. In October of 1969, the two presented their conclusions and, to the consternation of the AEC, they (1) supported Sternglass's contention that low-level exposure did have public health consequences,

although they noted that Sternglass had overestimated the effects and consequences, and (2) recommended that AEC (and Federal Radiation Council)
exposure limits be reduced by a factor of 10. They argued that the AEC's
exposure policy was based on a belief that a small increase in public health risk
was worth the public benefits of nuclear power, but that the AEC had grossly
underestimated those risks. They further argued that, when calculated realistically, the known risks plus the uncertainties outweighed the benefits and
required more stringent standards.[k]

The Commission was subsequently charged with forcing Gofman to resign
and then with systematically stripping Tamplin of his budget and research staff
as retribution for publication of their findings. The charge was denied, but the
publicity and circumstantial evidence did nothing to inspire public confidence in
the impartiality of the AEC.

Whatever the merits of the arguments, it became clear to the lay observer
that respected members of the scientific community held differing opinions on
the safety of nuclear reactors. It also became increasingly apparent that the real
question was not whether reactors were "safe" in the colloquial sense, but rather
if their benefits outweighed their costs (or risks), and on that question there
seemed to be no consensus.

As the scientific debate raged and the issue was formally raised by
intervenors at licensing proceedings, the State of Minnesota independently
adopted radioactive waste disposal standards for normal operating effluent that
were 100-fold more stringent than those required by the AEC. Minnesota did
this by writing standards into the state waste disposal permit requirements,
arguing that the technology to meet the tough standards was well in hand and
that no industrial enterprise had the right to contaminate beyond necessity.[13]

Northern States Power Company needed the permit before it could
discharge stack gases or cooling water from its 548 MWe Monticello reactor. But
not wanting either to pay for the extra radiation suppression equipment or to let
the precedent of a state's setting independent standards go unchecked, the
utility chose to sue on the grounds that the AEC had preemptive authority over
emissions standards. Thus in August of 1969 the states' rights battle was joined.
The State of Minnesota carried the legal battle all the way to the Supreme Court,
and in 1972 the High Court ruled that the federal government had exclusive
authority under the doctrine of preemption to regulate the construction and
operation of nuclear power plants, including the levels of radioactive effluent

---

kTamplin and Gofman estimated the increased incidence of cancer attributable to ionizing
radiation was some twenty to thirty times the AEC estimates. See Gofman and Tamplin,
Low Dose Radiation and Cancer; cf. The Institute of Electrical and Electronic Engineers,
Inc., Transactions on Nuclear Science 1 (1970). Also see Gofman, Gofman, Tamplin, and
Kovich, *Radiation as an Environmental Hazard,* Proceedings of the 24th Annual Symposium
on Fundamental Cancer Research, Environment and Cancer. Also see Gofman and Tamplin,
"Nuclear Power, Technology and Environmental Law," *Environmental Law,* Winter 1971,
pp. 57-73.

discharged.[14] Minnesota's effort to set standards independent of federal standards was struck down.

The course of the battle over radioactive discharge levels was similar to that over thermal levels. Once the issue had been identified by the scientific community and legitimized by some government support, intervenors began to question the adequacy of AEC standards. The issue was raised in sixteen different licensing hearings between 1966 and 1971. Although intervenors received little satisfaction from the AEC or from the Atomic Safety and Licensing Boards, they were again able to bargain independently with the utilities for installation of systems to reduce effluent gas, to delay its release and reduce its radioactivity, and to filter it. In combination, these systems could reduce radioactive operating emissions virtually to zero, well below the Gofman and Tamplin recommendation.

As the public awareness of the controversy and the availability of remedies grew, the AEC and its standards came under increasing fire. In 1970, Congressman Chet Holifield (Democrat/California), a major Joint Committee figure, had to fight a serious congressional effort to transfer responsibility for radiation emissions and exposure levels to the newly formed Environmental Protection Agency. Maryland and Illinois were threatening to join Minnesota in setting independent standards. Both Tamplin and Gofman were making numerous public appearances on behalf of their position and finding new support in the scientific community. By April of 1970 the mounting pressure prompted the Commission to issue a careful but vaguely worded formal proposed rule change advising utilities to keep radioactive effluents "as far below limits . . . as practicable."[15] The proposed rule change represented a stronger statement of intent than previous informal requests by the Commission, but it stopped short of actually changing the standards. The Commission wanted to maintain its flexibility.[16]

Environmental groups persisted in forcing utilities to meet their "zero release" target until finally, in May of 1971, the AEC issued a clearly stated set of "guides" specifying substantially reduced operating emission levels as meeting the "as low as practicable" criterion. The AEC's new levels were lower than those recommended by Tamplin and Gofman, but they were design objectives, not standards, and therefore could be breached occasionally.[1] Although the issue continued to be debated and the guides were revised over the next two years, the AEC's action did mute the criticism of its performance.

Both thermal pollution and low-level radiation releases were easily understood by the layman. They also stemmed from normal plant operations and were therefore both obvious and certain. It took the intervenors more time to see and to develop the expertise to exploit the more technical and probabilistic issues of emergency core cooling and the fuel cycle.

[1]The conclusions reached in the National Academy of Science's two-year study on the biological effects of radiation (concluded in May 1972) supported the Tamplin-Gofman findings and praised the new AEC guides. See *Nucleonics Week*, November 5, 1972.

The Emergency Core Cooling System

The Ergen Committee Report went largely unnoticed by the environmentalists. It was issued somewhat before they took a serious interest in nuclear power, and only a limited number of copies were made available.[17] To appreciate and exploit the committee's reservations, the intervenors needed a firm grasp of several complex technical issues, including why the emergency core cooling system was necessary, how it should function, and how to assess the probability that it would be needed. And in all likelihood they did not have that expertise in the late 1960s, since support from the scientific community was yet to come.

In spite of the serious nature of the committee's reservations, the issue of plant safety under accident conditions and the adequacy of the emergency core cooling system did not come to public view again until 1971. In May of that year the AEC made public the results of the "semiscale" emergency core cooling system test conducted at the Idaho test facility, and, as stated earlier, the AEC was sufficiently alarmed by these unexpected results to issue emergency Interim Acceptance Criteria the following month.

Only after the test results and the AEC's concern were well publicized did the environmentalists realize the potential of the issue. Also, by 1971 they had sufficient scientific support to exploit it. The adequacy of the emergency core cooling system had not been a major issue in any licensing hearing before 1971, but it was raised in virtually all construction permit and operating license interventions thereafter. And the AEC devoted considerable time and resources to resolving the issue over the next four years (see chapter 9 for a detailed discussion of the AEC's efforts).

Waste Disposal

The issue of radioactive waste disposal followed much the same pattern. After 1964, when Congress allowed private ownership of nuclear fuels, responsibility for the fuel cycle fell to the utilities. But because the technologies for fuel reprocessing and waste disposal were not yet in hand, the utilities stored most spent fuel at the reactor sites, awaiting some acceptable alternative. In the late 1960s, the AEC developed a process for solidifying radioactive wastes, theoretically enabling them to be permanently buried in certain stable geographic areas.

In 1970 the Division of Waste Management selected such a site in an ancient salt mine bed under Lyons, Kansas, and began an experimental waste disposal project. But the project encountered immediate local opposition spearheaded by Kansas Congressman Joe Skubitz, with technical support from the director of the Kansas Geological Survey, William W. Hambleton. Hambleton questioned the assumptions used to predict the behavior of such a salt mass when it contained quantities of very hot radioactive material. There was no empirical data supporting the AEC assumptions, and in the absence of good data, he argued, it was foolhardy to bury high-level wastes *irretrievably* as an experiment. He also questioned the geological stability of the site, noting that a neighboring salt mine had been closed because of water seepage. As the battle between the state and the AEC raged, previously unidentified gas bore holes were discovered in the bed. Those bore holes provided potential access for water, and by late 1971 the AEC was forced to abandon the project.

As in the case of the emergency core cooling system, once the issue became public and the AEC's position was questioned by a reputable authority, the environmentalists perceived and exploited their advantage. The question of waste disposal was raised at a number of hearings after 1971, and it became a priority issue in the environmentalists' educational campaign against nuclear power.

Regulation 1968-1971: An Assessment

The late 1960s definitely marked a new period of public concern for the quality of life, of public participation in the decision process, and of public skepticism of government's intentions. It was a time that surely tested both the information base for regulatory decisions and the acceptability of the risk levels implicit in AEC standards. They were found wanting.

In the case of exposure to low-level radioactivity, the AEC had clear primary responsibility for determining what exposures constituted a threat to the public health and safety and for controlling power plant emissions so exposures could not reach that level. Rather than developing its own research capability to produce or verify the information upon which its standards and regulations were based, the Commission simply relied on the National Committee on Radiation Protection's standards, an unofficial group with very limited research capabilities. Consequently the Commission was slow to incorporate changes in scientific thinking into its decisions and, when challenged, it had great difficulty defending its position. Its lack of aggressive safety R&D left the AEC at a similar disadvantage in the waste storage and emergency core cooling system controversies (see chapter 6).

At the same time, the AEC seemed singularly unresponsive to signals that public attitudes toward uncertainty and risk were rapidly changing. The Commission, although under great pressure, dodged responsibility for thermal effluents and was obdurate in its unwillingness to tighten controls on low-level radioactive emissions when the technology was available, the costs were relatively modest, and the outcry growing. The Commission demonstrated the same insensitivity to public attitudes when it insisted upon pursuing the Lyons, Kansas, waste disposal experiment to its conclusion.

In early 1971, the AEC responded to the intervenors' attacks with a proposed amendment to the Atomic Energy Act that would have (1) limited public intervention to an early mandatory hearing to be held at the point of site authorization, (2) permitted intervention at the construction permit hearing *only* if an unresolved public health or safety issue could be raised, and (3) prohibited intervention at the operating license hearing, where the delays were most costly and the intervenors exercised the most leverage. Even the Joint Committee on Atomic Energy could not support the AEC in such a blatant effort to bar the public from the licensing process, and the amendment was dropped in subcommittee.

Why the AEC seemed so blind to the consequences of jeopardizing public confidence in its expertise and concern for the public safety is not clear. In each case, the specifics of the decision vary. But the Commission's consistent determination to refuse perfectly feasible compromises reflects an arrogance and failure to understand its own political identity. Quite possibly the Commission's eagerness to see the reactor succeed caused it to be more responsive to industry and shut its eyes to other pressures.

Events of the late 1960s and early 1970s demonstrated the basically political nature of safety regulation. The regulatory body plays the role of "expert" as it sifts and evaluates information identifying the nature and level of the risk. But when it *sets* standards defining what risks are acceptable, it acts as a representative of the public will, since those choices are subjective value judgments. To function in this latter capacity the regulatory body needs to acknowledge its political role, remaining sensitive and responsive to the public will. If it does not, in a democratic system, its opponents can and will find alternative access to the decision process. If the regulatory authority is sufficiently out of touch, it might be reconstituted or power might be transferred, as was proposed in several congressional measures introduced between 1969 and 1973 and finally accomplished in 1974.[m]

Notes

1. JCAE, *Hearings on AEC Authorizing Legislation, FY 1973,* 972, p. 1071.

2. Harold P. Green, "Risk-Benefit-Calculus in Safety Determinations," *George Washington Law Review* 43:3 (March 1975), 792.

3. *Nucleonics Week,* March 5, 1970.

4. *Calvert Cliffs Coordinating Committee* v. *AEC,* 449 F.2d 1109 (D.C. Dir. 1971).

[m]The AEC was abolished and its developmental and regulatory functions were passed on to the Energy Research and Development Agency and the Nuclear Regulatory Commission, respectively.

5. *Public Law No. 92-500,* October 18, 1972.

6. Michael Goodman, "National Radiation Health Standards—A Study in Scientific Decisionmaking," *Atomic Energy Law Journal* 6:3 (Fall 1964).

7. Goodman, "National Radiation Health Standards. . .," p. 236.

8. Ibid., p. 240.

9. Ibid., p. 242.

10. Ibid., p. 244.

11. See PL 86-373.

12. Joint Committee on Atomic Energy, *Hearings on Environmental Effects of Producing Electric Power,* 91st Congress, 2nd Session, Part 1 (1970), p. 1588.

13. Richard S. Lewis, *The Nuclear Power Rebellion* (New York: Viking Press, 1972), p. 122.

14. *Northern States Power* v. *AEC,* 447 F.2d 1143 (8th Cir., 1971), *aff'd. per curiam* 405 U.S. 1035 (1972).

15. *Nucleonics Week,* April 16, 1970.

16. Gofman, Tamplin, "Nuclear Power. . .," *Environmental Law,* p. 71.

17. Gillette, "Nuclear Safety. . .," *Science,* p. 972.

8

Regulation and Costs

Dr. Okrent, ... you are kind of laying down these academic rules for people coming in for licenses, and you said they ought to be conservative in their design. I'm just wondering how fast we are going to progress if the committee [ACRS] is insisting upon this retention of conservatism all the time.[1]

—Representative Craig Hosmer, 1967

A regulatory authority of the kind we are examining is charged with guaranteeing that the risks associated with a product or its use are held to an acceptable level. To carry out its mandate, the authority typically sets standards and then adopts procedures by which it can review the product or its use to insure the standards are met. It must also develop a process that does not interfere unduly with the normal functioning of the marketplace. Neither the standards nor the review process should deprive society of a product that does *not* pose an unacceptable risk. An assessment of the degree to which the AEC's regulation of commercial reactors interfered with the adoption of nuclear power cannot be drawn easily from a narrative of events. But because it is an important question, it warrants attention. This chapter, therefore, represents a departure from the chronological flow for the purpose of evaluating how much and in what ways regulation deterred commercial adoption of the nuclear reactor.

When costs and schedules grew dramatically during the late 1960s, the AEC was held largely responsible. Both the nuclear industry and prominent numbers of the Joint Committee complained that regulatory procedures were increasing permit and construction times and, as a consequence, threatening parts of the United States with power shortages.[a] Moreover, they agreed that these delays, coupled with increasingly stringent standards, were largely responsible for the escalating plant costs that threatened the competitive position of nuclear power.

Plant completion time and costs did indeed rise rapidly over the years from 1966 through 1971. A nonturnkey plant ordered in 1966 was completed in about seven years and cost about $445 (1976 dollars) per KWe capacity.[2] A plant ordered in 1975 will probably take about ten years to complete and will cost about $700 (1975 dollars) per KWe capacity.[3] But, contrary to the many

[a]Joint Committee Chairman Holifield and Representative Hosmer complained to AEC Chairman Seaborg that the licensing procedure was hopelessly muddled and prolonged. They even suggested that the 1954 Atomic Energy Act be amended to limit intervention. Their suggestion was taken when the AEC proposed such a measure the following summer (1971). Joint Committee on Atomic Energy, *Hearings on Licensing Procedure . . .* , 1971, p. 1572.

allegations of the time, the evidence indicates that the AEC was not solely responsible for the increases.

Licensing and Construction Times

The time it takes to get a typical nuclear plant on line breaks into roughly three periods. First, there is the period between application for a construction permit and the beginning of the construction permit hearing. During this period the utility submits its application, and the regulatory staff and the ACRS conduct their safety evaluations. At the same time, the utility holds the plant site, develops its plant design, and may elect to do a minimal amount of site preparation.

Second, there is the time from the beginning of the hearing to issuance of the construction permit: for the most part, intervenors control the length of this period. The utility's main expenses during these two periods included manpower and land holding and amount to only a small fraction of total project costs.

The third period is the construction period, defined as the time between issuance of a construction permit and approval of an operating license. While regulations required another more detailed safety evaluation of the final design and actual plant as well as an operating license hearing, utilities generally applied for operating licenses well before plants were finished, and the regulatory staff almost always finished the safety evaluations and hearings before the utility completed construction.b The construction period constituted about 75 percent of the time from application to operation.[4] And since the bulk of the investment was made during this period, delays during construction, especially late in the period, might be expected to be particularly costly.

Regulatory requirements could affect the time it took to complete a proposed plant in two ways. First, the review process and hearings could be drawn out or extended past completion of actual construction, thereby delaying plant operation and the generation of revenues. Second, regulatory requirements could dictate increasingly complex safety systems and plant designs. And if these requirements changed while construction was in progress, engineering changes (ratcheting) might be required, thereby delaying completion of the plant.c

There can be no doubt that completing the procedural requirements for a

---

bIn a few cases the Commission issued a temporary license that allowed operation but restricted power output to some small percent of capacity, thereby effectively lengthening the construction period. But for the purposes of this discussion, a plant will be defined as on line when any operating license has been issued.

cCompleted plants might also be required to backfit to comply with changing standards, also a time-consuming and expensive process.

construction permit took an increasing amount of time (see figure 8-1). According to William Mooz, author of one of the very few quantitative studies of the subject, the average time it took to get a construction permit (review time plus hearing time) increased linearly between 1966 and 1971 at the rate of about five months per year (after controlling for location and other plant characteristics). Table 8-1 shows the licensing period separated into review and hearing times.

The review time increased during this period for many reasons. The number of applications, both construction permit and operating license, grew. Operating licenses took longer and were more costly, and the ratio of operating license to construction permit applications also grew.[d] But the AEC did not add a commensurate number to its review staff because there was a government-wide effort to hold down new hiring at that time (see figure 8-2).

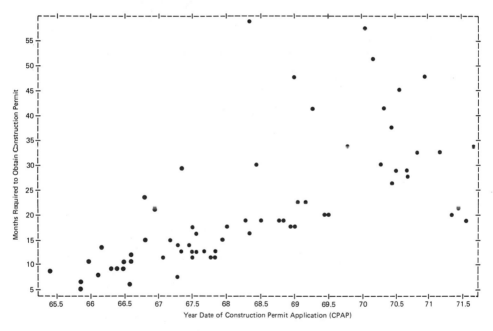

Source: William Mooz, *Cost Analysis of Light Water Reactor Power Plants,* fig. 2, p. 9. Santa Monica: The Rand Corporation, R-2304-DOE, June 1978.

Note: Some of the data points overlap.

Figure 8-1. Construction Permit Times for all 93 Plants That Applied for Construction Permits Prior to September 1971

[d]Operating licenses reviews take almost half again as much time as construction permit reviews. Because the design is complete, the regulatory staff has considerably more to review. In addition, the construction and design changes must be monitored.

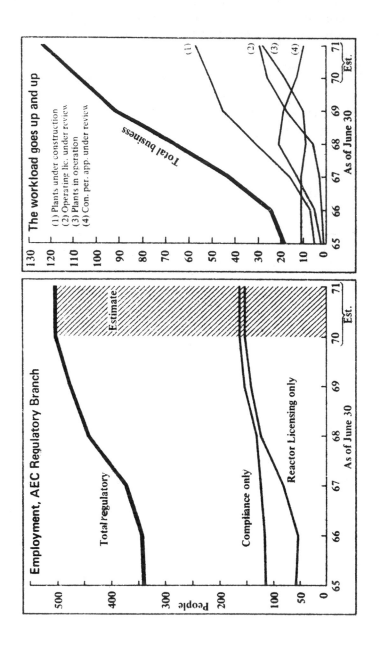

Source: *Nucleonics Week*, March 12, 1970. Reprinted from *Nucleonics Week* with permission of McGraw-Hill, Inc.

Figure 8-2. Staffing and Workload of the AEC Regulatory Personnel, 1965-1970

Table 8-1

Average Days of Processing Time at Various Stages of AEC
Applications for Construction Permits for Nuclear Power Plants

	1966[a]	1967[b]	1968[c]	1969[d]	1970[e]
1. Application to notice of hearing	214.0	251.7	344.1	411.8	460.8
2. Notice of hearing to hearing	37.0	36.4	42.9	47.4	41.1
3. Beginning to end of hearing[f]	2.0	7.7	6.4	11.7	53.7[g]
4. End of hearing to initial decision	12.3	34.5	40.3	36.3	52.0[h]
Total: Application to initial decision	265.3	330.3	433.7	507.1	607.6

Source: Emory N. Ellis and James H. Johnston, *Licensing of Nuclear Power Plants by the Atomic Energy Commission,* Administrative Conference of the United States, April 1971, p. 21.

[a]Figures for 1966 include four cases in which the hearing was started in that year.

[b]Figures for 1967 include eleven cases in which the hearing was started in that year.

[c]Figures for 1968 include seventeen cases in which the hearing was started in that year.

[d]Figures for 1969 include seven cases in which the hearing was started in that year.

[e]Figures for 1970 include twelve cases in which the hearing was started in that year.

[f]Hearings are frequently interrupted; this does not indicate the number of hearing days.

[g]Includes time elapsed up until February 1, 1970, for cases in which hearing commenced in 1970 but for which Initial Decision has not been issued to date.

[h]Does not include data for five cases for which hearing has not been concluded.

While the ratio of professional staff to cases under review decreased, there was no trend toward standardization: designs continued to change with each application, and the size and technical complexity of the plants increased quite markedly. Moreover, the new safety questions being raised by the scientific community (see chapter 6) forced regulatory staff reviews to be increasingly thorough.

For its part, the AEC staff blamed most of the review delays on the applicants themselves. Between 1963 and 1970, twenty-eight individual construction permits and nine operating license reviews were conducted that took more than twelve months to complete. The causes of these delays, as recorded by the regulatory staff, are indicated in table 8-2.

Hearing times varied considerably from plant to plant. If the permit or license went uncontested, the hearings were short—on the order of a couple of days—and contributed insignificantly to the length of the project. If, however, they were contested, the delay could be long indeed. But a 1972 AEC study of seventy-five representative plants found that only nine were delayed solely by licensing interventions. Of the seventy-five cases only thirty-six experienced intervention at all, and of these thirty-six, eighteen suffered no delay. Nine were delayed by intervenors and nine were delayed by a mixture of intervention and other problems (technical, labor, availability of equipment, and so on). In a later

Table 8-2
Causes of Delay—Construction Permit Staff Review

Cause	Instances
Applicant-caused	25
Incomplete application	(10)
Slow response to questions	(1)
Site or design switch	(3)
Construction exemption (special procedure)	(11)
New or unusual safety or site features	9
AEC-caused	1
Defects or construction delays (for operating license only)	6

Source: JCAE, *Hearings on Licensing Procedure and Related Legislation,*
pp. 565-571. Washington, D.C.: Government Printing Office, June and July,
1971. The total instances exceed the total causes because one delay was
occasionally attributed to several causes.

Federal Power Commission study of twenty-eight nuclear plants scheduled to
begin operation in 1973 but delayed, intervention (described as legal challenges)
contributed to the delay in only four cases (see table 8-3), and then accounted
for only 3.5 percent of the total time delays for the twenty-eight cases.

That such delays tended to be modest is not surprising. Hearings constituted
only a small fraction of the licensing process and an even smaller fraction of the
entire design and construction process.

In sum, increases in the review and hearing times were attributable to a
number of causes. Plants were growing larger and more complex. Staff work-
loads were increasing. Applicants themselves caused delays by filing incomplete
applications and not allowing for clearly predictable application procedures. And
although intervenors may not have been the direct cause of much delay, they, no
doubt, made the regulatory staff considerably more cautious and conservative.

Construction time (the time between issuance of a construction permit and
an operating license) for those plants granted permits between 1966 and 1972
increased and is plotted in figure 8-3. Mooz's data show about half of the
observed variance in construction time to be associated with increasing plant
size, the experience of the equipment manufacturer, and the date when
construction was begun. This analysis fails to explain the other half of the
increase.[5]

For each additional 100 MWe of plant capacity, construction time increased, on average, a little more than four months. Therefore, on average, a
1,000 MWe unit would take twenty months longer to build than an equivalent

Table 8-3
Causes of Schedule Delays in 28 Nuclear Plants Scheduled for
1973 Operation[a]

	Number of Plants Affected	Plant Months of Delay
Poor productivity of labor	16	84
Late delivery of major equipment	9	68
Change in regulatory requirements	8	23
Equipment component failure	6	15
Strikes of construction labor	5	18
Shortage of construction labor	5	18
Legal challenges	4	9
Strike of factory labor	4	5
Rescheduling of associated facilities	1	12
Weather	1	9

Source: JCAE, *Hearings on Nuclear Powerplant Siting and Licensing*, p. 571. Washington, D.C.: Government Printing Office, 1974.

[a]Note some of the numbers here differ from those above, because Ellis and Johnston (see table 8-1) use a different data set. They examine those plants that *held their construction permit hearings* in 1966-1970, not those that applied for a construction permit. Theirs is, therefore, a somewhat earlier group.

500 MWe unit. At the same time, each doubling of the number of reactors built by a particular engineering firm is associated with an 8 to 9 percent reduction in construction time. Conversely, apparent difficulties with one particular supplier, Babcock and Wilcox, increased the average construction time of all plants built during the six-year period by thirteen months. And finally, Mooz's results show that, on average, construction times lengthened at the rate of four months for every year later the plant was begun, after correcting for the effects of size, engineer, and manufacturer.[e] That is, on average, a plant granted a construction permit in 1970 could be expected to take eight months longer to complete than the same plant granted a permit in 1968.

The increases in construction time that could be associated with the date construction began (the date the construction permit was issued) and that portion that could not be explained by Mooz's analysis most certainly reflect a variety of influences. But most delays could be tied to the fact that commercial scale nuclear plants were bought in quantity off the drawing board with little prototyping and no operating experience. Under those circumstances, the unexpected should be expected. Early problems stemmed from unanticipated

[e]The cited effects of the various characteristics on construction time are based on regressions of construction times on these characteristics as reported in Mooz, p. 18.

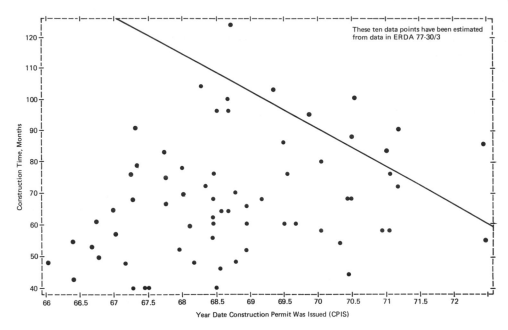

Source: William Mooz, *Cost Analysis of Light Water Reactor Plants,* fig. 4, p. 14. Santa Monica: The Rand Corporation, R-2304-DOE, June 1978.

Note: Some of the data points overlap.

Figure 8-3. Construction Time Data for 55 Completed Plants, and Ten Plants for Which the Data Have Been Estimated

bottlenecks in the manufacture and installation of major components. Ironically, the major culprit was the turbine/generator, not part of the nuclear steam supply system at all. Because the nuclear reactors produced low quality (cooler) steam, nuclear plants needed larger turbines (200' x 50' as compared with the 170' x 30' turbine required for a similar fossil plant). Predictably, but unexpectedly, the larger turbine occupied more manufacturing space and thus cut U.S. production capacity back considerably. Producers of other components found themselves in similar situations.[6]

Labor difficulties also contributed to longer construction schedules. Because the technology was new, unions fought the inevitable jurisdictional battles. For the same reason, appropriately skilled labor was in short supply and expensive.

Because utilities were unfamiliar with the technology, they were poor consumers and unable to demand the quality necessary for reliable performance. Reactors constructed in the late 1960s and early 1970s consistently had problems with pressure vessels, electrical systems, and vibrations, all resulting from poor prototyping and quality assurance. These defects caused construction delays, retrofitting, and delays in plants with similar designs or components. The fact that nuclear technology was new and unusually demanding was not fully appreciated by the utilities during those years.[7]

Although supply problems, labor difficulties, and poor quality assurance probably account for some measure of the lengthening construction time, increasingly demanding safety requirements must also have been responsible for a major portion of the change.

Costs

According to Mooz's data, plant capital costs (in constant 1976 dollars) increased at the substantial rate of $117 (1976 dollars) per installed KWe per year for plants issued permits between 1966 and 1972. Figure 8-4 shows the cost per installed KWe plotted against the permit date for that plant. Note that while there is a definite upward trend, costs for individual plants vary widely around this trend, suggesting there may have been a variety of reasons for and means of controlling cost increases.

While the real cost increases were sizable, the *perceived* increases were substantially greater, because the early cost estimates had been well below actual costs for early plants. Plants ordered in the mid-1960s cost about twice what the

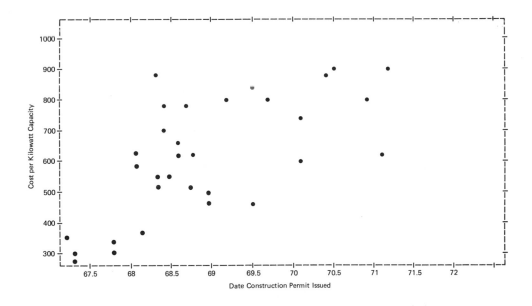

Source: William Mooz, private communication.

Figure 8-4. Costs and Permit Dates

utilities originally estimated. And these optimistic estimates coupled with actual cost increases left observers with the impression of extraordinary cost overruns and instability in the production system. The increases were, therefore, viewed with heightened alarm.

Even with the benefit of hindsight the nuclear community continues to disagree over the causes of the cost increases. As table 8-4 indicates, there is a strong tendency to blame "the other guy."

In his analysis of cost data, Mooz estimates the effects of a variety of plant characteristics on costs with somewhat surprising results. He explores how cost might be affected by a number of variables including time required to get a construction permit, construction time, size, cooling towers, location, manufacturer of unit, location of another unit on the site, and time construction was begun. Table 8-5 shows the factors that do appear to affect construction costs and the degree to which they can be associated with increases and decreases.[f]

Together, these characteristics explain about 70 percent of the increase in nuclear plant capital costs between 1966 and 1972.[8] However, it must be remembered that while four of the characteristics are directly tied to plant costs, the most influential characteristic, the date construction began, is obviously a

Table 8-4
Causes of Cost Increases

A Utility's View	The AEC's View
Changes in project scope	Poor productivity of labor
Environmental studies	Late delivery of major equipment
Safety questions	Change in regulatory requirements
AEC question/answer requirements	Equipment component failure
Additional radiation shielding	Construction labor strikes
Reactor proection system changes	Shortage of construction labor
Radiation monitoring system changes	Legal challenges
Engineered safeguards	Factory labor strikes
Fuel cask decontamination and transportation	Rescheduling of associated facilities
Mechanical draft cooling towers	Weather
Additional off-gas treatment	
Sharp rise in labor, materials and interest	
Delay of more than three years	

Source: A.C. Bupp, "The Economics of Nuclear Power," *Technology Review,* February 1975, p. 22, where Bupp notes: "Electric utilities and the Atomic Energy Commission hold perfectly contradictory views of nuclear reactor cost increases, each essentially blaming the other. The utility view represented here is that of Northeast Utilities, and the A.E.C. view is taken from a report in *Chemical and Engineering News,* November 26, 1973." Reprinted from *Technology Review,* edited at the Massachusetts Institute of Technology; copyright 1975 by the Alumni Association of the Massachusetts Institute of Technology.

[f]These numbers are taken from the summary of Mooz, *Cost Analysis,* based on his regression equations of cost on these characteristics.

Table 8-5
Factors Increasing and Decreasing Construction Costs

Characteristic	Change in Cost
Date construction begun	+$140/KWe/year
Located in Northeast	+$130/KWe
Cooling towers	+$ 90/KWe
Learning (per doubling of plants built)	−10% plant cost
Plant size (per 1 MWe increase in capacity)	−$.22/KWe

proxy for a group of subtler factors that the analysis could not uncover. And most probably the new safety requirements adopted by AEC between 1966 and 1972 contributed substantially to the cost increases reflected in Mooz's time variable.

However, the measure of a regulatory authority's competence is not whether it increases costs but whether it increases them "unnecessarily." To increase costs *unnecessarily,* the authority would have to make requirements that either exceed the public's demand for safety or increase costs without increasing the margin of safety. Or the authority could increase product costs indirectly through procedural inefficiencies that resulted in delays. The evidence cited in chapter 7 suggests that AEC requirements between 1966 and 1972 did not exceed the public's demand for safety. To the contrary, they fell short. Whether regulatory requirements imposed during these years reduced the risks inherent in nuclear power is a technical judgment and therefore falls outside the scope of this analysis. That leaves us with the final question concerning the possible costs AEC procedural delays may have imposed on the technology. Interestingly, Mooz's data indicate that delays (represented by increases in construction time) have no measurable effect on plant costs.[9] This fact would suggest that utilities and architect/engineers are able somehow to cope with delays caused by strikes, delivery lags, and probably also by lengthening review periods and changing design standards.

In sum, then, it seems clear that we cannot say what did cause the cost increases, although changing design requirements certainly played a role. What we can say is that since increasing construction times did not seem to affect plant costs, the procedural requirements that may have lengthened licensing and construction times should not have increased costs.

Notes

1. Joint Committee on Atomic Energy, *Hearings on Licensing and Regulation of Nuclear Power.*
2. William E. Mooz, *Cost Analysis of Light Water, Reactor Power Plants,*

R-2304-DOE, The Rand Corporation, Santa Monica, California, 1978, Table 15, p. 59.

3. Montgomery and Quirk, *Cost Escalations in Nuclear Power*, 1976, unpublished, p. 19.

4. Mooz, *Cost Analysis*, p. 22.

5. Ibid., p. 18.

6. W.E. Hoehn, *The Economics of Nuclear Reactors for Power and Desalting*, RM-5227-PR/ISA, The Rand Corporation, Santa Monica, California, 1967.

7. This discussion draws heavily on an AEC Staff Report by Roger Lagassie.

8. Mooz, *Cost Analysis*, p. 31.

9. Ibid., pp. vi, 41.

Coming of Age

Experience has shown that resolution of the [safety] issue has only been possible when the conclusion is supported "beyond a reasonable doubt" by the available evidence.[1]   —Fred C. Finlayson, 1975

Between 1963 and 1971 several historical threads converged, seriously jeopardizing the AEC's regulatory control over nuclear power. The rapid pace of commercialization, persisting technological change, and the new requirements imposed by environmentalists and environmental legislation overloaded the licensing process to the breaking point. At the same time, two apparently serious safety problems emerged—emergency core cooling and waste disposal—for which the AEC had no apparent solution. Reassessment of the emergency core cooling system threatened to require major design changes and possibly the backfitting of reactors already in service. The AEC's regulatory machinery seemed unable to cope with either the volume or the nature of the workload.

The opportunity for change presented itself in July of 1971 when Chairman Seaborg resigned after serving as chairman of the Commission for ten years. President Nixon appointed James Schlesinger, an economist and three-year veteran of the Bureau of the Budget, to succeed Seaborg as chairman. Schlesinger was a forceful personality who enjoyed strong presidential support. He took immediate control, shrewdly choosing to move Commission headquarters from Washington, D.C., to Germantown, Maryland, where, for a number of years, the agency staff had been enjoying the independence that comes with absentee management. As *Science* observed shortly after his appointment, "clearly Schlesinger's interests bear less in the direction of research and the production of transplutonium elements [where Seaborg's interests lay] and more in the direction of management techniques, environmental affairs, and weaponry."[2]

His personality, his political base, and his managerial talent permitted Schlesinger to bring about major changes in AEC policy. He came to the task with a very clear sense of his objectives. He was, in fact, a strong supporter of commercial nuclear power, but he understood that ultimately the future of the industry depended on its acceptance by the public. So he determined to improve the AEC's reputation as a defender of the public interest by strengthening its regulatory function and giving it more independence from industry. But he also saw the need to rationalize and speed up the licensing process. To help him with the restructuring, Schlesinger had William Doub, formerly chairman of the

Maryland Public Services Commission, appointed a fellow commissioner. Later he secured the appointment of Manning Muntzing to replace Harold Price as Director of Regulation.

Schlesinger's first opportunity to prove his commitment to these objectives came only two days after he took office. On July 23, 1971, the D.C. Court of Appeals held in favor of the Calvert Cliffs Coordinating Committee, requiring the AEC to meet all NEPA requirements retroactively. In spite of sharp opposition from industry and dire predictions that the ruling would totally obstruct the licensing process and ultimately leave the nation short of power, Schlesinger reversed earlier AEC policy and announced that the agency would accept the court decision without appeal.

In a speech before the Atomic Industrial Forum and the American Nuclear Society's joint annual banquet held in late October of 1971, Schlesinger reinforced his dramatic departure from past policy with a calculated and explicit statement of his rationale and goals. He noted that "the pace of achievement [in the nuclear industry] will depend heavily on two provisos: first, provision of a safe, reliable product; second, achievement of public confidence in that product."[3] He then laid responsibility for the safety of the product squarely in the lap of industry, stressing the importance of sound engineering and quality assurance.

He described the AEC's role as that of a referee serving the public interest. More specifically, as he saw it, it was the AEC's responsibility to conduct its business in an efficient manner so as not to be the source of delays and to avoid changing the rules for other than sound reasons. By implication, Schlesinger believed that if he could credibly establish the AEC's independence as a regulatory body, disassociated from the nuclear industry and its well-being, he would have gone a long way toward regaining the public confidence necessary for the continued diffusion of the technology.

The nuclear industry generally disagreed with Schlesinger's reasoning, advocating less regulation and less public participation in power generation decisions. The utilities were certain that abiding by the Calvert Cliffs decision would mean more intervention and further increases in hearing times. But industry's opposition was halfhearted, because it was as clear to the utilities and manufacturers as to the AEC and Congress that some change was imperative.

Licensing

The large number of applications, changes in regulatory responsibility, the lack of standards, and the *ad hoc* review process all combined to create a growing bottleneck in the licensing process. The utilities were particularly concerned by the steady rate of increase in the time it took to get a construction permit. As Manning Muntzing noted several years after he became Director of Licensing,

It is surprising that the quality of the AEC's technical review of applications is as good as it has been. It has been done in the past without a comprehensive study of standards, without a standard approach to data collection, without a standard review plan, and on a project rather than a functional basis. A price has been paid in time taken to reach decisions.[4]

Muntzing was not alone in his surprise. *Nucleonics Week* reported that Schlesinger and Doub were appalled by the apparent confusion and lack of organization they found in the licensing process.[5]

Schlesinger believed that assuring the expeditious review of license applications was the AEC's primary responsibility toward the nuclear industry. He therefore set as his target reducing the construction permit review to twelve months and the operating license review to sixteen months. And he intended the operating license to be issued upon completion of the plant.[6] The confusion of reorganization combined with the additional environmental review requirements imposed by the Calvert Cliffs decision forced the AEC to suspend licensing for most of 1972. During this period, Schlesinger, Doub, and Manning undertook a five-pronged overhaul of the licensing process. The regulatory organization was streamlined to make licensing more efficient and more responsive to the public interest. The regulatory staff got new leadership and was strengthened with more safety and environmental experts, and new tools for management analysis and control were introduced.[7]

Staff and Management

In April of 1972, Muntzing announced the reorganization and consolidation of his regulatory staff into three directorates: Licensing, Regulatory Standards, and Operations (Compliance). Conceptually, these directorates represented the three prerequisites of efficient control of the technology and harked back to the reorganization of the regulatory staff in 1961. Doub and Muntzing also added 375 members to the regulatory staff, bringing it to 1,393 full-time employees. They gave particular attention to the areas of reactor safety and environmental impact. At the same time, the Licensing Directorate adopted a formal schedule for application review based on critical path planning.

Applications

In keeping with a process developed in the early days when it licensed only a few experimental and development reactors each year, the AEC's license application requirements varied depending on the predilection of the project monitor. In 1972 the AEC standardized the application process, requiring preliminary safety analysis reports and enforcing clear guidelines on the preparation and content of application documents. This step was intended to help clarify the review criteria, set uniform reviews, and reduce the long review "question and answer" period. And if an "incomplete" application was filed, the licensing staff simply returned it with no further review scheduled, and the applicant was forced to return with his completed application to "the end of the line."

Regulatory Standards

During the 1960s, considerable attention had been paid the question of adequate regulatory standards. Industry wanted the AEC to settle on some comprehensive and fixed sets of criteria for licensing. The AEC had long acknowledged the need, but given the rapid pace of technological evolution, the lack of adequate R&D support, and the regulatory staff's traditional case-by-case approach to licensing, little had been accomplished. Schlesinger, Doub, and Muntzing seemed acutely aware that a sound set of standards was the basis of an efficient and defensible licensing process. Standards could define the information required in applications, provide inspection and enforcement criteria, limit contested issues in hearings, provide models and methodology for assessing levels of safety, safeguards, and environmental protection, provide the basis for standardizing plant design, and help inform the public. The men made this a priority objective.

First they upgraded the agency's own ability to develop adequate standards. The Regulatory Standards staff was increased, and in June of 1973 the directorate received increased funding to contract for outside technical assistance. Initially the emphasis was on drafting guides that could be effective immediately. New guides were developed and old standards were clarified and tightened. These included seismic standards (1973), new emergency core cooling system criteria (1973), more conservative assumptions regarding fuel anomalies (1972), stiffer radiation emission guides (1973), and quality assurance standards. It was not that the AEC was necessarily demanding larger safety margins. Rather, the technology was still new, and research and operating experience continued to turn up new evidence that suggested the safety margins were not necessarily as great as had originally been subjectively presumed.

The AEC and representatives from the American National Standards Institute identified priority areas, and after a little more than a year's effort the institute had drafted tentative standards in most areas for AEC approval. Inevitably, industry complained as the new standards were applied, particularly when they affected plants retroactively, requiring backfitting or redesign during construction. Nonetheless, the new Commission enforced the revised standards without hesitation, while Schlesinger observed:

> It is not our responsibility . . . if a utility encounters unanticipated costs because of a failure to do its job properly, failure to comply with

the procedures, or because of a change in the law. We are sympathetic; we understand your problem, but it is your problem.[8]

Admittedly, some ratcheting was simply a product of overzealous project review. To combat the "better mousetrap syndrome," a new policy was adopted requiring a licensing project manager to get approval from top management before he could impose a design change on a project under review. The utilities could then appeal any changes with which they disagreed.[a] This technique proved moderately effective in reducing the number of late changes.

Hearing Procedures

The Commission made several changes to streamline the licensing process. A prehearing conference was instituted to reach early agreement on the issues to be contended, the hearing schedule, and the availability of information. New regulations then imposed time limits on various phases of the hearing process and limited the kinds of challenges allowed in both construction permit and operating license hearings. The AEC also promised to provide intervenors with broader access to documents and urged the industry to do likewise.

By 1973, both Schlesinger and Doub had also come to support legislation that would abolish the operating license hearing. They argued that safety problems had to be identified early in the design and construction of a plant, and that when a plant reached the final construction stage, it should and would eventually be operated. Therefore an operating license hearing was not a real decision point. Rather, it offered intervenors a unique opportunity to blackmail utilities, because the costs of delay were so high at that point.[9] However, environmentalists continued to oppose any proposal that might reduce their influence on licensing decisions and Congress was becoming increasingly skeptical of the AEC's ability to balance environmental and safety objectives against development goals. Therefore, the Commission proposal was even less politically viable in 1973 than it had been in 1971, when it was first proposed.

The Commission took the important step of isolating some of those generic questions that were common to many or all plants but had been contested at individual licensing hearings. Separate rule-making hearings were held for each of those issues and the questions became inadmissible at licensing hearings. Not only did the use of rule-making hearings lead to shorter licensing hearings, but at the outset they appeared to offer a better forum for attracting expert

---

[a]For an example of the effect of ratcheting, in February 1974 every plant under review experienced a delay. Of these, design changes delayed twenty-one plants for one or more months. Ten of the twenty-one delays were caused by AEC requirements; two because the ice condenser needed redesign, three because poorly placed steam lines jeopardized safety equipment and needed relocation, and five because revised seismic calculations required design upgrading. All were serious, not frivolous, matters. (JCAE, *Hearings on Nuclear Power Plant Siting and Licensing*, 1974.)

contributions and for considering the full implications of an issue (a full discussion follows).

Public Confidence

There can be no doubt that restoring public confidence in the AEC was one of Schlesinger's prime objectives. He said as much, and many agency procedural reforms of this period were adopted to further this end. First, the past policy of allowing early site preparation was reversed, and no ground preparation or building was allowed until the Environmental Impact Statement was complete and the construction permit had been issued. In 1974 Commission Doub testified that prohibiting early site preparation "was an important necessary step taken, among other reasons, to increase the credibility and effectiveness of the AEC's regulatory hearing process."[10] The AEC thus gave full support to the spirit as well as the letter of NEPA and more credibility to the objectivity of its own licensing process.[b]

Second, on Commission orders, the agency did make some effort, albeit very imperfect, to be candid and to make documents available to the public. Prior to 1971, the AEC (like most federal agencies) had been extremely reluctant to open any of its files to public scrutiny, arguing executive privilege and that it had an obligation to protect proprietary information. Intervenors generally had to use their rights of discovery in formal proceedings to get documents they believed were necessary to building their legal cases and to educating the public on the true risks of nuclear power. The AEC's reluctance to be more forthcoming certainly strengthened the intervenors' argument that the agency was trying to hide the facts. In 1971 Congress passed the Freedom of Information Act, giving the public access to most public documents. The new legislation in combination with the Commission's new policy improved public access to AEC documents, although intervenors continued to complain that it was not good enough.

Agency staff pursued the Commission's policy of greater candor and contact with its antagonists by encouraging representatives of the AEC to meet with intervenors on various issues. Later, AEC employees who differed with the agency's official position were allowed to testify at the emergency core cooling system hearings, and AEC witnesses were available for extensive cross-examination.[c]

Third, rather than cutting off debate, the Commission took pains to

[b]Then in 1974, once it was clear a new independent regulatory body would be formed, the AEC again reversed itself to allow early site preparation.

[c]Ironically, during these same years the AEC got its share of negative publicity as Ralph Nader forced ACRS documents into the public domain under the new Federal Advisory Committees Act.

preserve the opportunities for criticism and appeared to weigh the intervenor arguments more seriously when adopting new or revised standards. (Note, for instance, the generic hearings on the adequacy of the emergency core cooling system as described below.)

But intervenors remained skeptical. They had come to see the AEC as an agency deeply committed to the development and application of nuclear power, and they viewed its regulatory mission and the regulatory staff as very much subordinated to the agency's developmental role. The opportunities for more cordial interaction during this period did little to restore their confidence or to deter them from their efforts to have the AEC's regulatory function vested in a completely separate institution.

It would appear that as time wore on Schlesinger himself also became considerably more skeptical about the possibility of bridging the gap between the nuclear community and its opponents. Early in his tenure, he accommodated the demands of intervenors and moved to include them on a more equal footing in the decision processes. The prolonged and bitter contest over the emergency core cooling standards seemed to suggest to Schlesinger that there would be no "reasoning together," that nuclear power's major opponents were not quibbling over standards but rather fighting the introduction of the technology in any form and using the only arena open to them. Thus the format of the generic hearing was changed, and he moved to eliminate operating license hearings.

Effect on Licensing

Schlesinger's decision to live with the Calvert Cliffs decision had an immediate and dramatic effect on the application review process. During 1972, review activity was suspended while the regulatory staff organization and regulatory procedures were overhauled. The Commission issued its new Rules of Practice in August of 1972, but by the close of the year only one operating license and no construction permits had yet been issued, leaving a substantial backlog of applications for the following year.

But the delays were only temporary. By the end of 1974 the average construction permit review time had been reduced from the twenty-five months required in 1971 to fourteen months, a greatly increased workload notwithstanding (see figures 9-1 and 9-2). And once the Limited Work Authorization was reinstituted in 1974, the environmental review time of only twelve months became the critical path and constraining figure (see figure 9-3).

Hearings were generally abbreviated. Although a substantially higher percentage of the applications was contested (70 percent of the 115 applications reviewed between 1970 and 1974), the average number of days devoted to hearings declined (see table 9-1).

Because the generic hearings diverted environmentalist opposition on certain

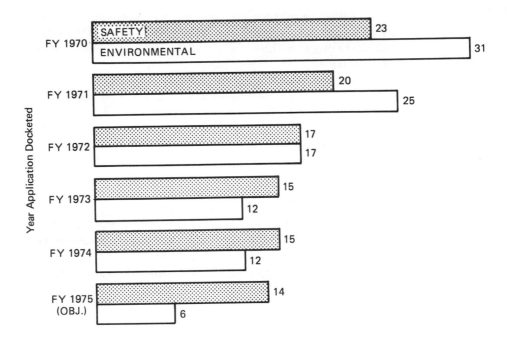

TOTAL YEARS OF OPERATION OF U.S. CENTRAL STATION NUCLEAR
POWER PLANTS OPERABLE AS OF FEBRUARY 1, 1972

aFrom date of docketing through completion of staff review.

Figure 9-1. Time Required for Construction Permit Reviews (Average in Months)a

key issues, the points of contention in individual hearings no longer provide a comprehensive means of identifying the issues that concerned them. During this period, the better-organized scientific and environmental groups concentrated their efforts on the generic hearings and a few well-publicized and precedent-setting cases to get maximum effect from their limited resources. They focused primarily on technical issues, especially the emergency core cooling system, but waste disposal, quality assurance, fuel rod anomalies, and the adequacy of AEC standards generally also became grounds for intervention. Local groups, again without access to good technical advice, were left to contest the ordinary license applications.

Major questions concerning the fuel cycle and plutonium recycling and safeguards were not generally pursued until the mid-1970s, again after they had first been raised by people within the nuclear establishment.[d]

Safety Problems

The 1972 reorganization of the regulatory staff and reform of the licensing process resulted in substantially more efficient and predictable regulatory

[d]There were a few exceptions where fuel cycle questions were the basis of interventions; these include the Big Rock Point, Ginna, and Vermont Yankee reactors.

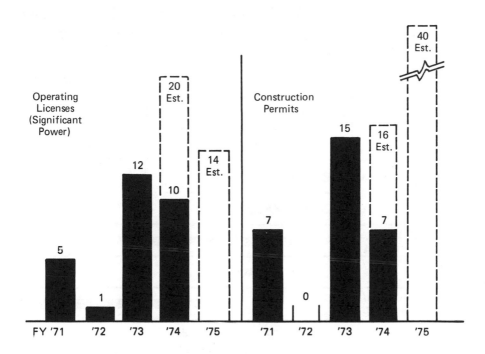

Source: JCAE, *Hearings on Nuclear Powerplant Siting and Licensing*, p. 349. Washington, D.C.
Government Printing Office, 1974.

Figure 9-2. Licensing of Power Reactors (Units)

procedures. But that change constituted only a modest improvement in the
overall regulatory process. If it wanted to defuse the arguments of its critics, the
Commission also needed to produce a convincing assessment of the hazards
nuclear power actually imposed and to set standards at levels that held the risks
"acceptably" low: both were very demanding assignments.

Table 9-1
Contested Hearings: 1970-1973

	1970-1972	*1973*
Number of hearings	33	19
Number contested	24 (73 percent)	13 (68 percent)
Range in months	4-28 (averages not available)	4-5 (average = 4)
Actual hearing days	3-70 (averages not available)	9-40 (average = 22)
Hearings lasting more than four months	18 (50 percent)	5 (25 percent)

Source: JCAE, *Hearings on Nuclear Power Plant Siting and Licensing*, p. 338.
Washington, D.C.: Government Printing Office, 1974.

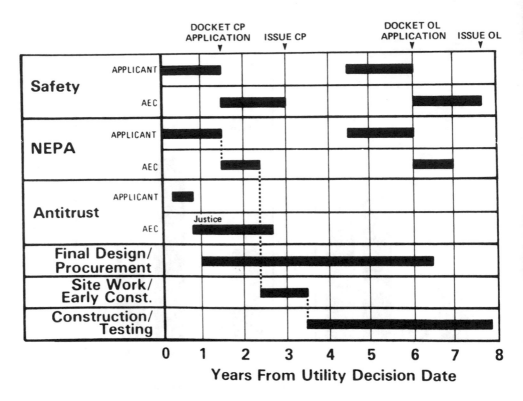

Source: JCAE, *Hearings on Nuclear Powerplant Siting and Licensing,* p. 358. Washington, D.C.:
Government Printing Office, 1974.

Figure 9-3. Reactor Design, Licensing, and Construction Cycle

Although nuclear power had been the subject of considerable regulatory
scrutiny since the mid-1950s, serious safety questions did not get public
attention until the early 1970s. As might have been expected, certain hazards
were not perceived until the technology took its final, tangible form. Other
hazards were perhaps thought possible, but they were not proven. Therefore, the
proponents of the technology were in a position to minimize and even suppress
the real hazard potential. And the regulatory staff, having to weigh a certain
economic benefit against a "possible" hazard, found itself under great pressure
to side with the proven good.[11]

By the early 1970s several changes combined to make reactor safety a more
sensitive issue. First, as noted above, increasing operating experience began
uncovering unexpected problems. As figure 9-4 shows, prior to 1971 the
industry only had the benefit of operating experience from a small number of

small plants. Between 1971 and 1974 plants in the 800 MWe range came on line. And these were followed by those in the 1,000 MWe range. Second, the Advisory Committee on Reactor Safeguards continued to put pressure on the Commission to investigate those questions the committee considered important, including possible failures leading to core meltdown, pressure vessel integrity, reactor fuel behavior, and sabotage. And third, the increasing scientific sophistication of both the intervenors and the public enabled opponents of nuclear power to exploit an even broader range of hazards inherent in the use of the technology. This capability forced the AEC to deal with questions more promptly, especially those concerning the fuel cycle and use of plutonium.

However, two underlying factors particularly threatened the Commission's ability to resolve the safety questions. First, the actual risk imposed by a reactor or, more broadly, by the introduction of a commercial nuclear technology, was a subject of great uncertainty.

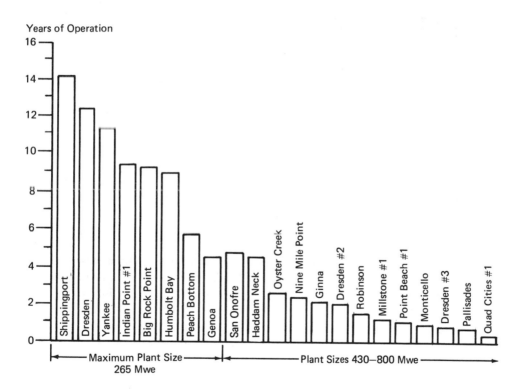

Source: JCAE, *Hearings on AEC Authorizing Legislation, FY 1973, 1972*, p. 1137. Washington, D.C.: Government Printing Office.

Figure 9-4. Total Years of Operation of U.S. Central Station Nuclear Power Plants Operable as of February 1, 1972

The answers to some questions, for instance the adequacy of the emergency core cooling system or the effects of low-level radiation exposure, could in theory be experimentally determined. But, in fact, either the research had not been done by agency choice, as in the case of the emergency core cooling system, or it could not be done in the time or on the scale required, given reasonable budget constraints, as in the case of the effects of exposure to low-level radiation. Because the effects of low-level radiation appear to be limited and only apparent some time after exposure, demonstration of those effects required research on large experimental populations over very long periods. Other safety questions centered on the reliability of institutions and therefore could not be resolved experimentally. The issues of radioactive waste disposal and of plutonium recycling fell into this category. And there was always the ultimate uncertainty introduced by the human element—saboteurs and fallible plant engineers and operators.

Second, the AEC had lost much of the public trust it enjoyed in the early 1960s. This change probably reflected a more widespread change in the public's attitude toward government. But the fact that the Commission and regulatory staff had consistently ignored very real public concerns over the environmental impact of nuclear plants and the consequences of exposure to low-level radiation certainly contributed to public suspicions. Growing public distrust forced the AEC to prove, in increasing detail, that the technology was safe. Given the uncertainties, the AEC could only offer an opinion.

Of the three major safety issues that surfaced prior to 1972, the Commission managed to resolve only one. By adopting a set of standards for normal radioactive effluents that met the most conservative public demands, the Commission was able to lay that issue to rest. Emergency core cooling and waste disposal standards became central topics in the nuclear debate of the early 1970s, and they were joined by several new issues.

Emergency Core Cooling

The adequacy of the emergency core cooling system remained the most vexing issue. In the *Interim Acceptance Criteria for Emergency Core Cooling Systems for Light Water Power Reactors* adopted by the Commission in June of 1971, the Commission noted that it would "consider holding an informal public rule-making hearing on this interim policy statement."[12] The ensuing controversy over those criteria prompted the Commission to order a generic hearing in January of 1972. And the hearing made clear the depth of disagreement over the adequacy of the criteria within the scientific community and even within the AEC.[13]

The hearing ran from January 1972 through July of 1973. Numerous technical witnesses from industry, from environmental groups, and from the

AEC hotly contested the appropriateness of the 1971 criteria. But after one and a half years, 135 acrimonious days of hearings, and over 22,000 pages of transcribed testimony, no concensus emerged. Industry believed the criteria were too conservative; the environmentalists believed they were too weak. Alvin Weinberg, head of Oak Ridge National Laboratory, publicly told Schlesinger that he had reservations about the adequacy of the emergency core cooling system.[14] Employees of the national laboratories with similar reservations anonymously sent supporting confidential evidence under plain wrapper to the environmentalists' coalition for use in the emergency core cooling system rule-making hearings.[15]

The AEC regulatory staff, author of the Interim Acceptance Criteria, heard the testimony, and the range of opinion and implicit uncertainties apparently persuaded the staff that its original recommendations were not sufficiently conservative. It therefore recommended to the Commission that the Commission adopt substantially more stringent criteria. In December of 1973, the Commission approved the regulatory staff's new criteria. The new criteria required many operating plants to reduce their power output until they complied with the new standards.[16]

There still were not sufficient test data in several important areas to prove the new criteria were based on conservative assumptions. The evidence supporting the analytical models used, the treatment of hot channel flow, and the assumed distribution of injected water (in pressurized water reactors) was either weak or incomplete. Moreover, although new emergency core cooling system research was given priority status after 1973, it emphasized "separate effects" tests rather than systems level performance, and many critics believed that simply extrapolating from the former could not provide the basis for defensible regulatory decisions.[17] As Weinberg noted in a letter to Schlesinger in 1972,

> As you know, much of our trust in the ECCS depends on the reliability of complex codes. It seems to me—when the consequences of failure are serious—then the ability of the codes to arrive at a conservative prediction must be verified in experiments of complexity and scale approaching these of the system being calculated. I therefore believe that serious consideration should be given first to cross-checking different codes and then to verifying ECCS computations by experiments on large scale and, if necessary, on full scale. This is expensive, but there is precedent for such experimentation—for example, in the full-scale tests on COMET and on nuclear weapons.[18]

Fuel Cycle Closure and Safeguards

Fuel cycle questions were largely ignored during the first two decades of nuclear power development. The Commission had concentrated its regulatory as well as its development efforts on the safety of the reactor itself. The first fuel cycle problem to get public recognition was permanent nuclear waste disposal. Then the Commission's determined pursuit of the Lyons, Kansas, site in 1971 guaranteed that some of the more dramatic potential dangers of disposal (safeguarding a facility for the half-life of plutonium, possible seepage of radioactive materials into the ground water, and so on) would receive wide publicity.

Once the AEC was seen to be incautious in its search for a solution to a clearly serious problem, the environmentalists turned waste disposal into one of the major topics in the safety debate. Subsequent reports in 1973 of major leakage in various AEC waste storage areas bolstered their case.[19] By the end of 1974, there appeared to be a standoff. The AEC still believed its strategy of solidification and burial would provide safe disposal. Much of the public, led by the environmentalists, viewed waste disposal as a major unsolved problem.

A new issue, that of plutonium diversion, surfaced in the early 1970s. Because of plutonium's extreme toxicity and its suitability as a trigger explosive in a nuclear device, the decision to separate and reuse plutonium in the fuel cycle demanded careful consideration. While the debate within the scientific community was underway, the Liquid Metal Fast Breeder Reactor was being developed and commercial reactors were being sold on the assumption that within the next decade or two both the breeder reactor and fuel reprocessing would be providing plutonium fuel for the commercial reactor.

The first evidence that the use of plutonium might become a public issue came in March of 1973 when the environmentalists sued to shut down the Big Rock Point reactor, which was using recycled plutonium fuel without having filed an environmental impact report. At the same time, security regulations protecting all radioactive materials and reactors against theft or sabotage came under intense scrutiny: five separate private and government reports on the subject were issued between November of 1973 and October of 1974.[e] Inevitably the two questions became publicly joined: Could plutonium fuels be adequately safeguarded?

In an effort to answer the growing criticism of its announced intention to introduce plutonium into the nuclear reactor fuel cycle and as a prelude to carrying out that intention, the AEC, in August of 1974, published a draft report, "Generic Environmental Statement on the Use of Recycle for Mixed Oxide Fuel in LWRs" (Wash-1327). The report recommended the prompt reprocessing of spent fuels and the recycling of the extracted unburned uranium and plutonium. It maintained that although existing safeguards were inadequate, they could be acceptably improved if plutonium were part of the fuel cycle.

---

[e]See "Improvements Needed in the Program for the Protection of Special Nuclear Material," General Accounting Office, November 7, 1973; "Protecting Special Nuclear Material in Transit: Improvements Made and Existing Problems," General Accouning Office, April 12, 1974; GAO Survey of Plant Security Systems, October 16, 1974; AEC Regulatory Study, April 26, 1974; Theodore B. Taylor and Mason Woolrich, *Nuclear Thefts: Risks and Safeguards* (Cambridge, Mass.: Ballinger Publishing Co., 1974).

In response to the severe criticism leveled in the various safeguards reports, the AEC tightened its security regulations considerably. By the end of 1974, the safeguards issue, as it applied to existing commercial reactor facilities and conventional nuclear fuel, appeared to be resolved satisfactorily. However, there was no consensus on industry's ability to introduce plutonium into the fuel cycle without incurring "undue risk." That subject was rapidly becoming the focal point of the nuclear power debate.

Design Problems

In addition to these older problems, several new technical problems emerged as plants came on line and accumulated operating experience. In 1972, when the Ginna reactor near Rochester, New York, was refueled, some fuel rod anomalies were discovered. The fuel pellets inside the rods had become more dense and had settled, allowing the cladding to collapse. It was a change in rod structure and fuel distribution that could have led to local hot spots and therefore theoretically could have disrupted the operation of an emergency core cooling system. Subsequently, similar rods were found in several Westinghouse pressurized water reactors, both in Europe and the United States. In November 1972, the AEC required all plants to provide a revised safety analysis assuming fuel densification and temporarily to reduce allowable power levels for reactors whose safety margins dropped below acceptable levels.

For two years, fuel densification and what to do about it was a major safety issue. But by the end of 1974, fuel rod manufacturers were able to provide a new and more stable fuel, effectively defusing the controversy.

Another design deficiency came to light about the same time. An employee in Minnesota's Prairie Island plant anonymously wrote the Commission pointing out that, in plants of that design, the primary system steam pipe left the reactor and passed by auxiliary safety equipment en route to the generators.[20] Thus, if a pipe failed at that point, the safety equipment would also fail. And while perhaps less well publicized, the regulatory staff identified a number of other potential design problems in the early 1970s, as it gained operating experience. These problems included stress corrosion cracking in stainless steel, fracture toughness in irradiated steel, and the possibility a reactor might not automatically shut down with a rapid increase in the chain reaction.[21]

None of these problems proved serious. They were all technical and could be solved relatively simply with research and design changes. But, as Schlesinger noted to the Joint Committee on Atomic Energy when discussing fuel densification,

> It [the fuel densification problem] is an indication of a number of aspects of the operation of reactors that have been insufficiently investigated. . . . I do not believe that they are serious enough to cause

concern about the safety of those reactors, but I do believe that they are serious enough for us to ask questions about the adequacy of the preliminary programs. . . .[22]

Moreover, any evidence of unanticipated technical deficiencies only reinforced the public's growing suspicion that the AEC could not capably protect its interest.

Quality Assurance

While the issue of quality assurance did not attract much public attention, Schlesinger, the Advisory Committee on Reactor Safety, and Milton Shaw unanimously described it as the principal safety problem.[23] Clearly the ability of a reactor's design and components to perform as they were expected to perform was essential to plant safety. Nonetheless, cases of pipe cracking, pump failure, and valve failure recurred. The rapid growth in demand put a severe strain on the reactor industry's design and production capability, which probably prompted hastier production and the hiring of less skilled labor. The utilities were generally unfamiliar with the technology and therefore not in a position, as consumers, to exercise quality control.

Schlesinger viewed an improvement in quality assurance as essential to a sound safety program, but it was not until 1973 that the AEC attacked the problem in earnest. In June of that year, the Commission issued more stringent quality assurance rules. Then, in December, in a precedent-setting move, the regulatory staff refused to authorize fuel loading for the 850 MWe Arkansas-1 reactor, because its electrical and control systems did not meet quality standards. By mid-1974, the Commission was threatening to fine the utilities for sloppy quality control. But as Norman C. Rasmussen, director of the AEC's definitive risk assessment study, observed in 1974

> Probably one of the most serious issues that the intervenors [critics of nuclear power] can raise today, with good statistics to back up their case, is that the nuclear power plants have not performed with the degree of reliability we would expect from machines built with the care and attention to safety and reliability that we have so often claimed.[24]

The problem of quality assurance could not be neatly disposed of in the manner of fuel densification, nor could it be characterized as a persisting point of controversy in the nuclear safety debate. Increasing experience within the nuclear industry and a stabilizing growth rate may improve the quality and reliability of the design and components. But since the safety of a given plant depends on the quality of that plant's design and components, quality assurance is always a potential problem and must weigh heavily in the uncertainty inherent in any assessment of risk.

Defining Acceptable Risk

Unresolved safety issues exploited by increasingly sophisticated intervenors continued to unsettle the development climate through the early 1970s. To reduce the uncertainties for industry and maintain its own viability, the Commission began trying new ways to rebuild a national consensus supporting the commercial use of nuclear technology.

In the past, public confidence in the AEC and lack of concrete evidence to the contrary permitted the Commission simply to assert that nuclear generating technology was safe. And "safe" in the 1950s and 1960s meant "risk-free." During those years, the Commission adapted a paternalistic stance, choosing not to scare people with the public discussion of risks, control mechanisms, and the possible consequences of accidents or the misuse of nuclear material. When faced with public opposition, the Commission avoided candid discussion of risks and benefits and, instead, generally tried to *educate* people to the benefits of nuclear power.[f]

As confidence in the AEC diminished and real safety questions became a matter of public concern, the AEC shifted its strategy. It acknowledged that nuclear power imposed a certain risk on society and moved to get consensus on the degree of that risk and on whether or not it was acceptable.

Generic Hearings and Consensus Formation

The Commission explored several avenues for reaching agreement on those questions. As discussed above, the generic hearing seemed to hold great promise. It provided the occasion for an adversary proceeding that would raise all the questions and provide a forum for presenting all the evidence relating to a given safety issue. Then, on the basis of the hearing record, the Commission could identify what risks the particular hazard imposed and what degree of risk seemed acceptable to the participants.

Two rule-making hearings were held in 1972. The first, a review of existing guides governing normal radioactive plant effluent, dealt with a topic already resolved on the basis of very conservative assumptions. It lasted seventeen days. The second, an evaluation of the 1971 Interim Acceptance Criteria for the emergency core cooling system, reviewed standards based only on moderately

---

[f]See, for example, Ashley, *Nuclear Power Reactor Siting*, 1965. Many discussions reflect the AEC staff's belief that people were being misled by a tiny vocal minority and that the AEC and the nuclear industry needed to beef up their public information programs.

conservative assumptions. As described above, the hearing led to a tightening of standards for emergency core cooling systems. But the Advisory Committee on Reactor Safeguards, many reputable scientists, and the intervenors believed that even the more conservative criteria adopted after the hearings were not adequately based, and the subject needed more research.

Not only did the Commission find that no consensus had emerged from this generic hearing, but because the issue was raised in an adversary context, the positions of the participants became polarized before the AEC could develop more experimental data. Intervenors spent their time arguing procedural issues and trying to get documents from the AEC and utilities. The fact that information was not readily forthcoming, despite the Commission's avowed policy to the contrary, only encouraged existing suspicions of the AEC. At the same time, the tactics of the intervenors so antagonized Schlesinger that he limited the role of the intervenors in the regulatory process wherever possible after that.

The AEC held two more generic hearings, one on the nuclear fuel cycle and one on the transport and disposal of fissionable material. However, after its experience with the emergency core cooling system hearings, the Commission based its last two hearings on a legislative rather than a judicial model, denying right of discovery or cross-examination. The hearings lasted only a few days each, and neither contributed much to the resolution of the issues under investigation.

The intervenors rapidly grew disenchanted with what at first appeared to be a genuine opportunity to address the substantive, generic issues. They believed they were not getting an unbiased hearing, and once the legislative hearing model was adopted, they lost their chance to force information into the public domain. Generic hearings normally held in Washington, D.C., did not offer the publicity or educational opportunities of licensing hearings in local areas, and removal of the generic issues from the licensing hearing seriously limited the intervenors' negotiating leverage.

Risk Estimates

At the same time the Commission undertook generic hearings, it also sought defensible, objective information on the actual magnitude of risks the use of nuclear power imposed. In 1957, the AEC had contracted for an analysis (Wash-740) of the "maximum credible" accident, its likelihood and its consequences, preparatory to the government's assuming a large share of the accident liability under the Price-Anderson Act. But at that time no commercial plants were in operation, so design details and consequently conclusions were necessarily vague. In 1965, the study was repeated in anticipation of a ten-year extension of the act, although this time the results were withheld on the grounds that the public would not understand them and would be unduly alarmed.

The studies were based on the simplest of models and the barest of data, while the technology was changing very fast. Therefore, neither one contributed much to the AEC's ability to quantify or express the risk burden imposed by nuclear technology in later years. In lieu of any satisfactory analysis, the AEC rested its case on the meager operating experience of the first ten years and the subjective assessments of its staff and its advisory committees.

By 1971, the agency urgently needed some objective basis for proving the overall conservatism of its regulatory requirements. Also NEPA required that all reactor environmental impact statements include a comparative cost-benefit and risk-advantage analysis for the variety of steam supply system options available to a single plant. The AEC responded with some pioneering work in risk assessment.

Prompted by a request from the Joint Committee on Atomic Energy, the Commission ordered its staff to prepare a comprehensive safety analysis of the commercial power reactors to identify the safety issues, document the risks, and compare the risk imposed by a nuclear plant with other everyday risks.[25] The report, written by Shaw's office without contributions from AEC lab scientists or the Idaho safety research group, was published in 1973. Most observers considered it a "cover-up" of some key safety questions and a "whitewash" of the agency's safety research program. But the report did constitute the agency's first effort to estimate failure and accident probabilities.[26]

While the in-house study was underway, the Commission contracted with a group headed by Norman Rasmussen of MIT for a definitive analysis. Rasmussen's group proposed to evaluate reactor safety by calculating the probabilities and consequences of a broad range of accidents and deriving risk assessments. The AEC hoped that such a study could reassure the public that, even in combination, the emerging technical problems did not pose a serious threat. Second, the AEC hoped that this analytical approach would direct attention away from the consequences of the "worst possible accident" to the *average risk.*"

The Rasmussen study represented a seventy-man-year effort and cost $4 million. The group's draft report, *Reactor Safety Study: An Assessment of Accident Risks in U.S. Commercial Nuclear Plants*, was issued in August of 1974. It concluded that, contrary to widely held opinion, core meltdowns might *not* be rare events, but it also suggested that they would only rarely cause extensive damage. And, although the AEC's hopes that public attention would be shifted from the worst kinds of accident to the average risk were probably realized, it also happened that Rasmussen's study highlighted areas where the AEC's assumptions appeared to be less than conservative. The AEC had been particularly weak in assessing probabilities of external events and common mode failures, and ironically, the Rasmussen study led to some further tightening of the AEC criteria and guides.

The study seems to have provided the reactor safety debate with a concrete point of departure, but it too received a full measure of criticism. Critics argued that there could be no guarantee that any analysis has included all possible failure paths. They pointed out that in the Apollo program, where the same analytical techniques were used, about 20 percent of the ground test failures and 35 percent of the in-flight malfunctions had not been identified as "credible" prior to their occurrence. Moreover, these techniques had predicted a reliability of one failure per 10,000 missions for the fourth stage of the Apollo rocket, when the highest actual reliability achieved after years of testing was four failures per 100 simulated missions.[27]

Critics also argued that many of the assumptions used in the models (for instance, descriptions of human error, the flow of radioactive releases, evacuation characteristics, and so on) were wrong. Some say they were too conservative. Others say they were too optimistic. Critics further noted that the report's conclusions may be misleading, because they were products of average characteristics, and averages obscure higher consequence-lower probability outcomes. Given the complexity of the subject and the immaturity of the methodology, it is quite likely that even a revised Wash 1400 will not quiet the debate over reactor safety. Moreover, the risks inherent in other steps in the fuel cycle (reprocessing, use of plutonium, and waste disposal) remain unevaluated.[g]

Determining the risk burden imposed by commercial nuclear power was only the first step in defusing opposition to further use of the technology. The AEC realized it also had somehow to identify that level of risk it could defend as *acceptable*. Back in 1967, Representative Craig Hosmer had asked Dr. Peter Morris, Director of the Division of Reactor Licensing, how the AEC determined what level of risk was *acceptable*. Dr. Morris responded ambiguously that

> . . . the guidance starts with the participation by the Commission and the strong support by the Commission to the FRC, the President's Federal Radiation Council, and its philosophy with respect to accepting the risks of radiation. Further, the Commission issues policy statements, the Commissioners have made speeches, testimony before this committee is all available to us.
>
> As you discussed earlier, Mr. Price meets with the Commission regularly, and I associate with him quite regularly.[28]

As the public began challenging the AEC's judgments, the nuclear community sought some more objective basis for its determination of acceptability. In 1971, Chauncey Starr, then Dean of Engineering Sciences at UCLA, undertook a study comparing the health risks of nuclear power with those imposed by other forms of power generation.[29] The purpose of the study was to show nuclear power to be as safe or safer than other power sources and thereby

---

[g]Wash 1250 purports to assess risks from reprocessing and waste disposal. But since neither technology has been commercialized or used, we must assume those numbers are "guesstimates" of the crudest kind.

to ascertain its acceptability. In its 1973 safety study (Wash 1250), the AEC pursued Starr's line of reasoning by showing that the risks from nuclear power fell well below those imposed by a broad range of other commonly accepted activities.[30]

AEC efforts to redirect the debate to the relative safety of nuclear power probably contributed substantially to the public's understanding of the real issues. The public seemed to develop a growing appreciation of the facts that no technology was risk-free and that safety had its price. But nonetheless the public seemed unwilling to accept the underlying inference that nuclear power constituted an "acceptable risk," and opposition to nuclear power continued to grow during those years. Finally, for lack of any better means of determining acceptability the regulatory staff quite arbitrarily adopted a risk target of 1×10^{-7} (or a probability of less than 1 in 1 million) for major accidents.[h] That meant AEC standards should be such that if all reactors were built to its specifications, there would be only one major accident for each 1 million reactor-years of operation.

Methodologies for defining and calculating acceptable risk burdens for nuclear power grew considerably more sophisticated in the years following Wash 1250. But critics persuasively continued to cite major difficulties. Since the fuel cycle had yet to be closed and substantial uncertainty remained over the reliability of existing safety systems and over the long-term effects of low-level radiation, the risk burden imposed by nuclear power could not be established.

Furthermore, even if it could have been established and measured, analysts did not understand how to judge the acceptability of the risk. For instance, were there factors that bias the public's perception of the risk?[31] How should risks be weighed against benefits? If people were risk averse, how should the very high consequence but low probability events be assessed? And what distribution of risks over the population did the public perceive as *equitable*? How could risks and benefits be weighed when one group received the benefits while a different group was involuntarily subjected to the risks? How should risks imposed on future generations be counted or discounted? Answers to these analytical problems were few.

Controlling the Technology

At the same time the AEC moved to shift the safety debate to *relative* and *average* risks, it also took limited steps to get firmer control of the nuclear technology and its development.

---

[h]JCAE, *Hearings on Nuclear Reactor Safety*, 1973, p. 495. It was generally understood by the nuclear community that this was the AEC's target, though it was never formally adopted as such.

Standardization

Perhaps the most pronounced characteristic of power reactor technology during its first decade of commercial use was the rapidity with which it evolved. It was a constantly moving target, and the regulatory staff was always at least a step behind it. Schlesinger, Doub, and Muntzing believed that to gain shorter licensing times, get operating experience, answer safety questions, and reduce backfitting, the nuclear plant design had to be standardized.

In June of 1972, the AEC announced its intention to endorse and give preferential treatment to 'standardization designs,'' but with no definition of what constituted a standardized plant. The nuclear industry made no comment on the new policy. Two months later five utilities joined together and announced their intention to build identical plants. At the same time, the Atomic Industrial Forum expressed its reservations, noting that standardization must evolve out of the licensing regulations and procedures. Industry's response could be described as mixed, but some companies were clearly reluctant to freeze the nuclear technology at this point and to forego further economies of scale or customized designs suited to unusual sites or utility preferences.

In March 1973, the AEC announced it was ready to accept applications for standardized plants. A standardized plant was defined as: (1) a manufacturer and architect/engineer's *reference plant* in which the nuclear steam supply system was standard, although the balance of the plant and the site could be customized, (2) *duplicate plants*, with the several applications being considered together, and (3) a *license to manufacture* for identical plants on similar sites (for example, off-shore plants). Somewhat later a fourth option, the *replica plant* (one duplicating a design already licensed), was approved in response to demands from industry. A week after the original announcement, the Commission set the power limit for the standardized plant at 3,800 MWth or approximately 1,300 MWe.

By the end of 1974, all the reactor manufacturers, but only a few architect/engineers, had submitted standardization applications for review. Utilities had also filed standardization applications for thirty-nine reactor units representing only eight different nuclear steam supply system designs.[i] Progress toward standardization appeared to stem as much from the utilities' effort to cut design and construction costs as from their willingness to accommodate regulatory needs. Nonetheless, standardization won some converts.

Safety R&D

Once the adequacy of the emergency core cooling system came into public question, Shaw grudgingly gave the subject more attention.[j] And in 1973

[i]These were not necessarily applications for construction permits. Some were for design approval.

[j]For budget increases in this area see JCAE, *Hearings on Nuclear Reactor Safety*, 1973.

Schlesinger described the safety R&D program as now emphasizing design-oriented technology and large-scale testing to get definitive experimental data.[32] But it was too late. By this time Shaw had made too many enemies in his pursuit of safety as he chose to achieve it. In May 1973, Dixy Lee Ray, appointed to succeed Schlesinger only three months earlier, created a new Division of Reactor Safety Research to be headed by Herbert Kouts. Having lost the battle to retain control of the safety research program, Shaw then left the AEC.

Although the general manager and not the director of regulation continued to have authority over the new division, the research program's responsiveness to regulatory needs improved over the next year.[k] But that was hardly time enough to rebuild a program. Although allocations to safety research increased, the Loss of Fluid Test facility and the emergency core cooling system separate effects work continued to absorb most of the budget. Little of major consequence was accomplished before the AEC itself fell victim to reorganization.

Regulation 1971-1974: An Assessment

The year 1971 was clearly a watershed for the AEC. In the face of a growing energy shortage, nuclear regulatory problems had become a matter of serious concern to the administration and to the industry. The industry also argued that increasing licensing times and changing regulatory standards jeopardized the competitiveness of nuclear power. And at the same time, the increasing list of safety questions caused growing public concern reflected in a variety of state and congressional challenges to AEC decisions. Under new leadership, the Commission assessed the strengths and objectives of external political forces more realistically than it had before and concluded that the future well-being of both the AEC and the nuclear industry depended upon some major reforms. The licensing process had to be made more efficient. The AEC had to resolve satisfactorily some rather serious safety questions. And the agency had to regain public confidence. As a measure of the importance attached to regulatory problems during those years, two commissioners (including the chairman) gave the subject a major share of their attention.

Streamlining the licensing process proved to be well within the capabilities of modern management techniques. The regulatory staff, while making the licensing process more objective and orderly, of necessity had to assume a more formal, adversarial relationship with the nuclear industry than it had in the past. And even as it accepted expanded responsibilities under NEPA, the staff was able to reduce review and licensing times by almost one-half.

[k]Perhaps a good measure of the sorry state of reactor safety research was the fact that after only one and a half years of service, Herbert Kouts, the program's new director, received the AEC's highest honor, the Distinguished Service Award, for "his courage and determination in taking innovative measures to restore confidence in the AEC's reactor safety program." Herbert J.C. Kouts, "The Future of Reactor Safety," *Bulletin of the Atomic Scientists*, September 1975.

Finding acceptable solutions to the several serious safety questions that surfaced during the late 1960s and early 1970s proved more difficult. The Commission had two means by which it could resolve these issues. It could find technical solutions to the problems, or it could determine the real risks imposed by the problems in question (accident control, waste disposal, and plutonium diversion primarily) and find that those risks, once identified, were acceptable to the public.

In the case of fuel densification, the industry offered a technical solution. But in all other instances, neither industry nor the AEC thought technical alternatives were sufficiently promising to be worth the investment of major research monies. They also did not choose to pursue technical research that might shed light on the capabilities of safeguards or the consequences of serious types of accidents, thereby reducing the uncertainties surrounding the risk burden imposed by nuclear power.

Instead, the AEC made some effort to force a consensus on the definition of the risk and on what risk was "acceptable" through the vehicle of public hearings. At the same time the Commission tried to shift the focus away from individual safety problems to the *overall* risk imposed by nuclear power. The risk assessment research sponsored by the AEC showed that nuclear power imposed less risk than a number of commonly accepted activities. But because risk assessment analysis usually *averages* a broad variety of risks, it has the property of obscuring the rare but high risk event. While the new analyses did shift the focus of the safety debate somewhat, they were not sufficiently persuasive to earn broad acceptance.

While the Commission tried in several ways to build a consensus on how serious the major safety problems were and what solutions might be satisfactory, it could not. The AEC was reaping the harvest of the many earlier decisions to make safety R&D a low priority, to ignore fuel cycle questions, and to allow commercialization to proceed at an uncontrolled rate. Risks changed and grew, and the regulatory agency was left without enough information to act convincingly in the role of protector of the public interest.

The Commission was no more successful in restoring public confidence in its technical competence and willingness to defend the public at the expense of the industry. In spite of clear efforts to be open, accessible, and more candid in its discussion of risks and benefits, the AEC continued to receive sharp attacks from intervenor groups. The AEC learned a hard political lesson: if one enjoys the confidence of the people, it is hard to lose it, but once lost, it is hard to regain.

Epilogue

As early as 1960, observers had begun questioning the appropriateness of merging development and regulatory functions in the Atomic Energy Commission. In spite of strong AEC opposition, support for measures to divide the agency according to these functions grew over the following years. In the early 1970s, environmentalists also began questioning the advisability of perpetuating an agency responsible for developing a specific technology. They argued that all energy-producing technologies should be developed and evaluated competitively according to their efficiencies and environmental effects.

By 1974, public skepticism regarding the AEC's ability to regulate in the public's interest, plus a perceived "energy crisis" that promised to push the federal government into a major energy development program, combined to set the stage for the AEC's dissolution. In that year a major reorganization bill created the Energy Research and Development Administration (ERDA) and a separate Nuclear Regulatory Commission (NRC). ERDA added the AEC's development activities to a host of similar responsibilities for other technologies. The NRC then fell heir to the AEC's regulatory responsibilities.

The reorganization has not caused any major transformations under regulation. Although the NRC is out from under the shadow of the AEC's development wing, it is mostly staffed by those who staffed the AEC during its last years. It has inherited the same regulatory traditions, it faces the same technical problems and uncertainties that plagued its predecessor, and its basic approach to safety through engineered safeguards has not changed. Nor has anything in the external environment moved to force a clearer consensus on the choice between more energy and less risk to the public health and safety.

Notes

1. Fred C. Finlayson, 'View from the Outside," *Bulletin of the Atomic Scientists*, September 1975, p. 20.

2. "Seaborg Chairman for a Decade: AEC Grew Old and Grey," *Science*, January 14, 1972.

3. *Nucleonics Week*, October 21, 1977, pp. 4-6.

4. Joint Committee on Atomic Energy, *Hearings on Nuclear Reactor Safety*, 93rd Congress, 1st Session, 1973, p. 357.

5. *Nucleonics Week*, October 8, 1971, p. 1.

6. Joint Committee on Atomic Energy, *Hearings on AEC Authorizing Legislation, Fiscal Year 1974*, 93rd Congress, 1st Session, 1973, p. 30.

7. AEC, *Annual Report to Congress*, 1972, p. iii.

8. Schlesinger, speech before AIF/ANS meeting, as reported in *Nucleonics Week*, October 21, 1971.

9. JCAE, *Hearings on Nuclear Reactor Safety*, 1973, p. 314.

10. JCAE, *Hearings on Nuclear Powerplant Siting and Licensing*, 1974, p. 1152.

11. Harold Green, *Technology Assessment*, Raphael Kasper, ed., p. 27.

12. AEC Regulatory Staff, "Interim Acceptance Criteria . . . " 36 Fed. Reg. 1/247-11250 (June 29, 1971).

13. See Docket, RM-50-1.

14. *Nucleonics Week*, May 17, 1972.

15. Steven Ebbins and Raphael Kasper, *Citizens Groups and the Nuclear Power Controversy* (Cambridge, Mass.: MIT Press, 1974), p. 211.

16. *Nucleonics Week*, March 23, 1972.

17. Finlayson, "A View from the Outside," *Bulletin of the Atomic Scientists*, September 1975, pp. 23-24.

18. Letter to J.R. Schlesinger, AEC Chairman, from A.M. Weinberg, Director ORNL (February 9, 1974) (included in Hearing Record, DKT. RM-50-1), as cited in William B. Cottrell, "The ECCS Rule-Making Hearing," *Atomic Energy Law Journal* 16 (Winter 1975).

19. AEC, *Atomic Energy Programs, Volume I*, Washington, D.C., 1973, pp. 116-117.

20. R. Gillette, "Nuclear Safety: AEC Report Makes the Best of It," *Science,* January 21, 1973.

21. JCAE, *Hearings on Nuclear Reactor Safety*, 1973.

22. JCAE, *Hearings on AEC Authorizing Legislation, FY 1974*, p. 37.

23. JCAE, *Hearings on Nuclear Reactor Safety*, 1973.

24. Murphy, *The Nuclear Power Controversy*, p. 166, as quoted from *Combustion*, June 1974.

25. JCAE, *Hearings on Nuclear Reactor Safety*, 1973.

26. R. Gillette, "Nuclear Safety: AEC Report Makes the Best of It," *Science*, January 21, 1973.

27. Phillips, "Energy Report . . . " *National Journal*, February 1, 1975, p. 157.

28. JCAE, *Hearings on Licensing and Regulations of Nuclear Reactors*, 1962, pp. 201-202.

29. See C. Starr, M.A. Greenfield, and D.F. Hanskrecht, *Public Health Risks of Thermal Power Plants*, UCLA-Eng-7242, May, 1972.

30. *Wash 1250*, Chapter 6.

31. For a discussion of this phenomenon, see Paul Slovic and Baruch Fishoff, *How Safe Is Safe Enough?* Decision Research, Eugene, Oregon. Unpublished paper.

32. JCAE, *Hearings on AEC Authorizing Legislation*, FY 1974, p. 46.

10 Conclusions and Observations

Even in an ideal world of universally accepted objectives, unambiguous standards, and unlimited resources, regulation would remain an intrinsically subtle and complex process.[1] —Irvin C. Bupp and Jean-Claude Derian, 1975

The history of the Atomic Energy Commission has the overtones of a dramatic tragedy. In 1954 Congress mandated the Commission to develop commercial nuclear power and to insure that it was "safe." The Commission undertook the commercialization mission, striving to make the technology as safe as possible consistent with its development mandate.

During the early years when there were only a few, small reactors, the risks imposed by the technology were small and public confidence in the agency was high. The scientific community, the nuclear industry, and the AEC cooperated informally, and regulatory decision-making was also informal and personalized. Because the technology was so new and untried, the Commission applied very conservative safety criteria during these years.

In 1963, General Electric made the first commercial sale of a nuclear powered generating plant, and the number of commercial orders rose rapidly over the next five years. Both the AEC and the nuclear industry were buoyant with optimism. They both believed the new technology to be completely in hand. Consequently, the only significant safety research of that period was directed at proving the reactor could withstand the worst credible accident, the core-melt, without releasing fission products into the outside environment. With the anticipated empirical proof, the AEC was prepared to relax its conservative safety requirements. Because the number of license applications in process remained reasonably low and because the public remained fundamentally unaware that nuclear power imposed a major risk and had continuing confidence in the AEC, the regulatory process appeared to function smoothly through these years.

But dramatic changes were occurring by the end of the decade. Nuclear technology had entered a period of rapid change and wide commercial acceptance. The new reactors were larger, with greater power densities and longer-lived fuel. There were also to be more of them. These characteristics all combined to increase substantially the risks imposed by the technology. At the same time new and unanticipated safety questions emerged, further increasing the perceived risks. And finally public attitudes toward nuclear power and toward the AEC appeared to undergo a rapid and dramatic shift. In this new climate the AEC was severely challenged and found wanting.

Information Resources

To make informed decisions limiting the risks imposed by commercial nuclear plants, the AEC needed to know what hazards existed and how effective proposed safety measures were. The quality of information the Commission used determined the quality of its regulatory decisions; the kind and the source of the information dictated, to a lesser degree, the credibility of its decisions. While good information was very important to the quality of AEC decisions and to the credibility of the agency, getting good information could be very expensive. Thus the Commission was confronted with some hard choices.

The Commission had three main sources of information open to it: the nuclear industry, the scientific community, or AEC-managed research. It could require industry to supply whatever information it needed. Because reactor manufacturers employed a highly trained engineering staff and conducted an immense amount of research that illuminated safety issues in the normal course of development, they were an obvious resource. Using the industry to conduct safety research also had the advantage of placing the costs for the research at the door of the user, not the public at large. However, industry was clearly limited in its physical and financial capacity to handle major safety research programs. For instance, it was highly improbable that the nuclear industry could or would have undertaken any experiments on the scale of the STEP series. Moreover, if the subject was at all controversial, industry data and results were invariably suspect.

The scientific community at large offered a second source of information upon which the AEC could and did draw. Independent experts, some from industry, but many from universities and AEC labs, who worked in areas related to reactor safety were called upon, either as consultants or as members of blue-ribbon advisory commissions. Contributions from experts were not suspect as were those from industry and they were also very inexpensive. However, consultants and commissions were limited in the kind of information they could provide. They were called upon for advice, which meant summarizing and evaluating existing empirical evidence and offering subjective judgments. But they had no capacity for new research.

Finally, the AEC had the option of managing its own safety research program. Such a program could have taken a variety of shapes. It could have been manned entirely within the AEC, and, in fact, this was the model the Commission chose for its limited program. Alternatively, much of the research could have been contracted out to universities or other independent groups. Or the Commission could have required a major industrial safety research program, designed and monitored by the AEC (on the FDA model). An AEC-managed program had the greatest flexibility and the greatest capacity to produce objective, credible data upon which to base regulatory decisions. It also offered the possibility of searching out and dealing with new or unanticipated problems. However, an AEC program of any size was certain to be very costly.

Through most of its tenure, the Commission decisions reflected the judgment that the costs of an extensive AEC-managed program would outweigh its benefits. Although the AEC had an unusually strong safety research capability in its several large laboratories, test facilities, and well-trained professional staff, the Commission and the regulatory staff relied heavily on industry data for setting standards and evaluating plant designs. And as a check on industry, they sought advice or support from experts: the Advisory Committee on Reactor Safety, independent consultants, or, occasionally, blue-ribbon panels like the Ergen Committee.

Moreover, the Commission successfully resisted most efforts by the ACRS and others to persuade it to undertake major safety research projects, except when the projects were expected to contribute to the economic success of nuclear power, as in the case of the Loss of Fluid tests and the Power Burst Facility tests. It tacitly supported Shaw as he curtailed existing light water reactor safety research programs and redirected remaining efforts to quality assurance. And if the funds were not available to explore well-known problems, they certainly were not available to uncover new, unanticipated hazards. Only in its final years did the Commission reverse itself, placing the safety research program under new management and expanding the program. But it was too late.

While the generalizations that can be made from one case are limited, the evidence suggests that the quality of AEC regulatory decisions and the credibility of the agency both suffered from an overreliance on industrial safety research results and subjective expert judgment. Conversely, it appears that the AEC regulatory program could have benefited considerably from an aggressive safety research program managed by the Commission and probably under the direct control of the regulatory staff. Such a program could have allowed research to begin on issues as they first were raised by members of the technical community and, with luck, the Commission then would have had enough credible data to make acceptable decisions as the issues later became a subject of public concern. Uncertainty could have been reduced and credibility improved.

It would be foolhardy to argue that more and better information would have eliminated opposition to AEC regulatory decisions. However, resolution of the low-level operating emissions and fuel densification controversies suggest that conservative standards based on strong empirical evidence or on evidence that satisfies the bulk of the scientific community can lay an issue to rest.

It would appear, with the benefit of hindsight, that regulatory decisions could have benefited from more early research on low-level exposure, core-melt accidents, certain quality assurance problems, radioactive waste disposal, and

other fuel cycle questions including the risks associated with extracting plutonium from spent fuel. Since the agency already had the existing infrastructure
of a safety program, large, well-staffed laboratories, and major test facilities, a
strong research effort could easily have been mounted. And in fact it would
seem plausible to argue that if General Electric and Westinghouse could afford to
lose almost $1 billion between them on turnkey plants, the AEC could have
raised a large research budget through some form of tax on the reactor
manufacturers, thereby returning research costs to the users of the product.

Procedures

As a context within which to arrive at regulatory decisions, the AEC had to
establish procedures for setting, applying, and enforcing standards. Not only did
those procedures insure an orderly and predictable conduct of agency business,
but they had a bearing on the quality of its decision-making in several important
respects. Most notably, the kinds of procedures adopted determined both the
efficiency of the regulatory process and its reputation for fairness. They also
defined, to some degree, the quantity and kind of information that came to the
attention of AEC decision-makers.

More often than not, there will be a tension between the procedural
requirements for these several regulatory objectives. The tension was quite
apparent in the AEC's regulation of the commercial nuclear reactor. One can
think of any set of procedural requirements as falling on a continuum between
informal and private and formal and public. Informal, private decision-making
has certain characteristics. It is likely to be more efficient if there is considerable
uncertainty surrounding the factual basis for a decision, because decision-makers
need not resolve the uncertainties before making a decision. That decision, then,
reflects the best "mainstream" expert judgment. However, if decisions are made
privately, they cannot be defended as fair or equitable. Informal, private
decision-making would seem to be an appropriate model when uncertainties are
great but public risks are limited. It is then that decisions need to be based on
subjective judgments and equity is not an issue.

Formal decision-making based on a public record has a different set of
attributes. Foremost, it is perceived as equitable. The factual basis for decisions
is publicly established, and every point of view receives a hearing. Formal
procedures generally include public participation during hearings and therefore
elicit more information, particularly information of a less orthodox nature.
Hearings can also assure that decision-makers are at least exposed to a sampling
of public attitudes. When there is little uncertainty, formal procedures can also
be efficient in that they define regulatory standards and hurdles in advance,
thereby reducing time-consuming negotiations between the applicant and the
regulatory authority. At the same time, public participation can make them

extremely time-consuming and inefficient. Moreover, when there is a high degree of uncertainty regarding the factual basis of a decision, such procedures seem to highlight the uncertainties, making defensible decisions more difficult to reach. Hence these procedures seem most appropriate to situations where uncertainties are low and equity would demand that the decisions to incur risk be formal and public.

The regulatory history of commercial nuclear power is, in large measure, the story of the tension between the demand for rapid reactor commercialization which, in turn, demanded a formal, public decision-making and the high level of uncertainty regarding the risks which, in turn, prevented public decision-making. In early years when reactors were small and few in number, the Commission made its regulatory decisions informally and in the comfortable privacy of Commission meetings. But when the risk grew sufficiently to elicit a public challenge, Congress (always more sensitive to issues of equity) quickly forced the AEC to adopt formal, public licensing review procedures. So long as decisions went unchallenged, difficulties inherent in these procedures did not emerge. However, once opposition to nuclear power developed, formal, public review procedures offered opponents the opportunity to highlight the uncertainties underlying regulatory decisions and substantially reduce the AEC's credibility.

Other characteristics of the Commission's decision-making also seemed to erode its credibility. A major reason Congress required that public hearings be the basis of the AEC's decision-making process was to insure that the process was perceived as equitable. But because public participation in the licensing process began well after the close, informal working relationship between the AEC and the nuclear industry developed, the AEC was viewed as a biased spokesman for industry. Its reports and decisions invariably supported the applicant on the industry's position, and in time intervenors came to view the hearing process as a charade and a frustration.

The hearings also kept the AEC in contact with a segment of public opinion. While the intervenors did not *represent* public attitudes, their emergence signaled a change in public attitudes and their growing activity clearly reflected (as well as prompted) increasing public support. However, rather than viewing the intervenors as a reflection, albeit imperfect, of public opinion, the AEC and the Joint Committee preferred to view them as an isolated "vocal minority" to be frozen out of the process. Thus, AEC decisions soon failed to mirror public values and public confidence in the agency waned.

Pacing the Development and Use
of the Technology

Admiral Rickover pursued a very deliberate course of reactor development and application in the naval program. He took a relatively simple and reliable reactor design, used a standardized model in his submarines until he had good operating experience, and then slowly made development changes. Rickover's strategy of pacing development allowed him to gather good operating data, thereby reducing uncertainties and demonstrating the reliability of the equipment. By contrast, the AEC chose not to intervene in the free market development of commercial nuclear technology until very late in its developmental history. In the absence of intervention, utility preferences, perceived economies of scale, and improvements in engineering dictated customized and constantly changing designs. Consequently, there was no strong record of operating experience to use in resolving uncertainties and demonstrating reliability.

Pacing development and use of a technology is clearly a conservative strategy that has economic as well as safety implications. It may well mean that the public foregoes the benefits of the product for a number of years. Or it might mean that the cost of the product is inflated because producers of the technology cannot take full advantage of economies of scale or improvements in design. On the other hand, it allows the regulatory agency the opportunity to resolve major uncertainties before making judgments about the acceptability of risk, and could, therefore, avert later delays and uncertainties stemming from instability in the regulatory process. Such a strategy might also bolster public confidence in the regulatory process, because it would show that safety and not development was the more important objective. How a regulatory agency weighs these outcomes should depend primarily on the public values of the moment.

Some Observations

Without a doubt, nuclear power has a unique regulatory history. But that history provides a number of insights that apply to regulation in general and to safety regulation in particular.

It seems there must inevitably be a tension between the demand to make a new product commercially available and the demand to fully understand the implications of large-scale commercial use *before* it is used. How rapidly a potentially dangerous product should be brought into use is essentially a value judgment and the province of the political process. And regulation should be understood as a political process. Interest groups are central. Decisions must be responsive to public values and to changes in public values. And no matter how one tries to insulate the regulatory body or its decisions, there is always access through alternative political channels. In this case, opponents of AEC decisions had the option of appealing to other levels of government or to new congressional coalitions to get AEC decisions reversed. And even if their appeals were not successful, they could be sufficiently threatening to force the Commission to reverse itself.

The fact that early public participation in the regulatory process does not appear to guarantee all issues of concern to the public will be raised is another interesting insight that runs counter to the conventional wisdom. As the nuclear case has shown, the public is not necessarily interested in a technology still in the development stage. Nor is the public generally technically competent to raise issues before experts recognize them. And, finally, public concerns change. Therefore, to the degree that it has an obligation to protect the public's interest as it is defined by the public, a regulatory agency must remain flexible and provide multiple points of access.

A strong R&D capability would seem to be an essential element in a strong and independent regulatory program. A good research capability enables a regulatory body to develop information upon which to base regulatory standards and explore and resolve possible problems before they are encountered in operating plants. No regulatory body can rely on the industry to conduct this kind of research voluntarily because the industry's research would inevitably be suspect, and industry tends to be interested in a more limited scope of problem than is a regulatory agency.

Furthermore, industry cannot be expected to regulate itself. Any industry is naturally interested in minimizing malfunctions and accidents to avoid shutdowns and maintain the image of its product, but it must also take risks to cut costs and insure a competitive market edge. The nuclear power industry was willing to take risks in excess of those the public was willing to accept, as expressed either by the AEC's standards or by congressional reactions.

The industry believed that nuclear technology was well in hand by the early 1960s, lack of operating experience notwithstanding. Pushing hard for approval of urban and seismic area siting, industry pursued a rapid course of development with little experimental and developmental experience. Vendors and utilities bitterly fought backfitting, although it was the inevitable result of their own chosen course of development. While industry was arguing that nuclear power posed no appreciable risk to the public, the utilities and vendors themselves threatened to abandon reactor development and commercialization unless the federal government provided special liability coverage and limits. None of these facts should be surprising in that the nuclear industry was properly reflecting its own interests, which cannot be expected to be coincident with the public interest.

No doubt, when this case study is supported by similar examinations of different regulatory experiences some of these conclusions will be strengthened, while perhaps doubt will be cast on others. But what must be recognized is that regulation of these potentially high-risk products is a field ripe for research and of great policy interest.

Note

1. Irvin C. Bupp and Jean-Claude Derian, "The Nuclear Power Industry," *Commission on the Organization of the Government for the Conduct of Foreign Policy*, Appendix B. "The Management of Global Issues" (Washington, D.C.: Government Printing Office, Vol. I, 1975).

Appendix A
Summary of the
Regulatory Process

Standards

First, what constituted the "standards" an applicant for a license must meet if the license was to be granted in a timely fashion? The answer is not straightforward. There were, of course, the standards formally adopted by the Commission that clearly had to be met before a facility could be licensed. These included codes and performance criteria the Commission adopted over the years, as well as its Rules of Practice.[a] A second kind of standard, the guide or guideline, technically constituted a design or performance objective only *recommended* by the AEC regulatory staff. But in practice there was little distinction between the formally adopted standard and a guideline, since compliance with the latter offered a utility the best hope for expeditious licensing of a plant.

Safety Evaluation Reports offered another informal source of standards. The AEC's regulatory staff performed an extensive evaluation of the plant design for each construction permit applicant. These reports often pointed out deficiencies in the safety of certain design or component characteristics. While unofficial, the critiques generally became the basis for voluntary future design modifications, again because the applicant was anxious to avoid possible delays for redesign.

Enforcement practices were a fourth source of standards. Unless the standards being enforced were very vague, it was difficult to make them more conservative than originally intended. However, consistent lack of enforcement, or enforcement at less than the formal level, was a time-honored technique for amending regulations.

The citizens' groups that intervened in the licensing hearings eventually exercised enough leverage to impose some design standards of their choosing. The threat of protracted hearings and court battles that might cause operating delays gave intervenors the power to negotiate improved safety and environmental standards.

Who Sets Standards

The Atomic Energy Commission had paramount responsibility. The Commission adopted all formal rules, criteria, codes, and Rules of Practice. The AEC

---

[a]The key sections in the Rules are 10CFR50 (governing plant design and operation), 10CFR100 (specifying site considerations), and 10CFR20 (establishing permissible radiation exposure levels).

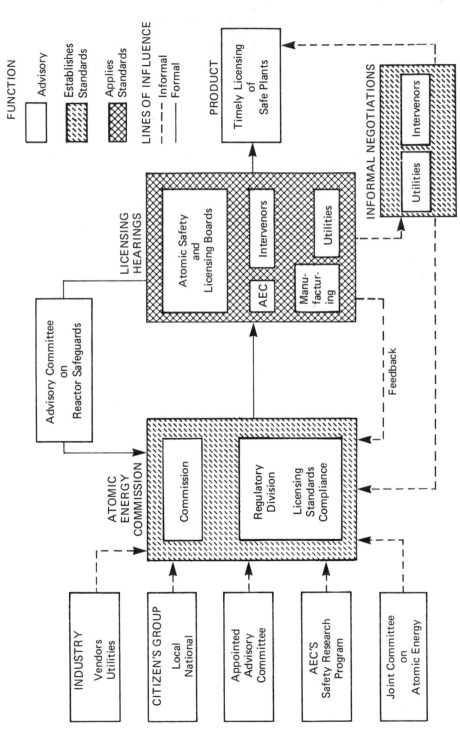

Figure A-1. The Regulatory Process

regulatory staff, particularly the Divisions of Standards and of Licensing, also played key roles.[b] Then, as described above, intervenor groups ultimately managed informally to set the standards governing radiation emission levels and waste water quality.

Who Influences

Formal interest groups representing manufacturers, the utilities, and environmentalists played their traditional roles of monitoring, recommending, and mobilizing support or opposition for decisions. In addition, several professional advisory groups exerted substantial influence. Foremost among them was the Advisory Committee on Reactor Safeguards, a statutory group of fifteen experts appointed by the AEC to advise it on all matters pertaining to reactor safety. In addition to monitoring the quality of the AEC safety program, the Advisory Committee on Reactor Safeguards reviewed each permit and license application. Because nuclear technology was new and empirical evidence on operating characteristics was scarce, the AEC frequently appointed special commissions from the scientific community and industry to make subjective assessments of particular problem areas.

A third and potentially powerful source of influence was the Joint Committee on Atomic Energy. Owing to its unique powers and close working relationship with the AEC, the Joint Committee could have influenced the Commission's regulatory decisions. In fact, although the committee expressed continuing interest in regulatory problems, it appears not to have fully exploited its power in this area.

Licensing hearings and hearings on general questions of reactor safety influenced AEC decisions. Hearings provided the Commission, although sometimes belatedly, with feedback information on whether there was agreement between the Commission and the public on what constituted undue risk.

In addition to interest group opinion, expert opinion, and feedback from hearings, the AEC had its own safety research program as a last source of information. This program offered the agency an opportunity to develop empirical data tailored to the requirements of safety analysis and standard-setting.

Issues

Many issues were fought out in the regulatory arena. Some were easily disposed of; others remained as visible impediments to the smooth diffusion of light water

[b]Over the years, reorganizations have changed the names of those divisions, but the basic responsibility for setting formal standards and for evaluating applications for permits and licenses has rested in separate divisions since 1961.

technology. For the most part, the conflicts centered on the adequacy of AEC standards governing normal radiation emission levels, seismic criteria, thermal discharge, adequate protection against the consequences of a serious accident, system and component standards (quality assurance), security of nuclear material, and radioactive waste disposal. Industry was also concerned about inefficiencies in the regulatory process.

The Process

Static descriptions imbue the actors and issues with a misleading stability. The regulatory process, of necessity, operated in a larger universe where concerns slowly but constantly changed. The concerns of the 1950s were not those of the 1970s. Not only did the external environment influence what issues might be at a given time, but the progress of technology prompted shifts of emphasis. To propose using an untested emergency cooling system in one 200 MWe reactor would not elicit the same public or regulatory response as proposing such a system for several 1,000 MWe plants. Pressure groups also changed. Their influence grew and waned with shifts in the political milieu. Most matured in organizational and lobbying sophistication over time, improving their ability to deal with complex processes and issues.

Also, implicit in real world interaction is the element of time lag. To act and get feedback takes time. To respond takes more time. And by the time the response is made, the original demands may well have escalated or shifted, leaving the demand still unsatisfied. This lag phenomenon was particularly apparent in the interactions between the hearing participants and the AEC on environmental and safety issues.

**Appendix B
Nuclear Facilities:
Characteristics and
Milestones**

Table B-1
Status of Central Station Nuclear Power Reactors, Significant Milestones

Project/Location	Owner	Cap. Net (MWe)	Type	NSSS/AE Contr.	Public Annc'd.	NSSS Contr. Award	CP/OL Applied	CP/OL Issued	Initial Crit.	First Elec.	Initial Design Power	Commercial Oper.
1 Shippingport Atomic Power Station (PA)	Duquesne Light Company and AEC	90	PWR	West. S&W	10/53	7/53	NA	NA	12/2/57	12/18/57	12/57	NA
2 Indian Point Station, Unit 1 (NY)	Consolidated Edison Co.	265	PWR	B&W O/Vit.	2/55	2/55	3/55 1/60	5/56 3/62	8/2/62	9/16/62	1/63	10/62
3 Dresden Nuclear Power Station, Unit 1 (IL)	Commonwealth Edison Co.	200	BWR	GE Bech.	4/55	7/55	3/55 6/58	5/56 9/59	10/15/59	4/15/60	6/60	8/60
4 Yankee Nuclear Power Station (MA)	Yankee Atomic Elec. Co.	175	PWR	West. S&W	4/55	6/56	7/56 9/59	11/57 7/60	8/19/60	11/10/60	1/61	2/61
5 Enrico Fermi Atomic Power Plant, Unit 1 (MI)*	Power Reactor Development Company	60.9	FBR	PRDC CA	4/55	3/57	1/56 7/61	8/56 5/63	8/23/63	8/5/66	10/70	a
6 Pathfinder Atomic Power Plant (SD)*	Northern States Power Company	58.5	BWR	AC PSE	2/57	5/57	3/59 6/62	5/60 3/64	3/24/64	7/25/66	9/67	b
7 Hallam Nuclear Power Facility (NB)*	Consumers Public Power District and AEC	75	SGR	AI Bech.	4/55	9/57	2/59 4/61	7/60 8/62	8/25/62	5/29/63	7/63c	11/63
8 Humboldt Bay Power Plant, Unit 3 (CA)*	Pacific Gas & Electric Company	65	BWR	GE Bech.	2/58	2/58	4/59 9/61	11/60 8/62	2/16/63	4/18/63	5/63	8/63
9 Elk River Nuclear Plant	Rural Cooperative Power Association and AEC	22	BWR	AC S&L	2/56	6/58	3/59 7/60	12/59 11/62	11/19/62	8/24/63	2/64	7/64d
10 Peach Bottom Atomic Pwr Station, Unit 1 (PA)*	Philadelphia Electric Company	40	HTGR	GAC Bech.	11/58	11/58	7/60 2/64	2/62 1/66	3/3/66	1/27/67	5/67	6/67
11 Carolinas-Virginia Tube Reactor (SC)*	Carolinas-Virginia Nuclear Power Associates, Inc.	17	HWR	West. S&W	8/57	1/59	7/59 1/62	5/60 11/62	3/30/63	12/18/63	9/65	3/64e
12 Piqua Nuclear Power Facility (OH)*	City of Piqua, Ohio and AEC	11.4	OMR	AI H&N	2/56	6/59	9/58 2/61	1/60 8/62	6/10/63	11/4/63	1/64	2/64f
13 Big Rock Point Nuclear Plant (MI)	Consumers Power Company of Michigan	75	BWR	GE Bech.	12/59	12/59	1/60 12/61	5/60 8/62	9/27/62	12/8/62	3/63	11/65
14 Bolling Nuclear Superheater Power Station (PR)*	Puerto Rico Water Resources Authority and AEC	16.5	BWR	Comb. J&M	6/58	1/60	12/59 2/62	7/60 4/64	4/13/64	8/14/64	9/65	g
15 Genoa Nuclear Generating Station (WI)*	Dairyland Power Cooperative	50	BWR	AC S&L	4/61	6/62	10/62 8/65	3/63 7/67	7/11/67	4/26/68	8/69	2/71
16 Haddam Neck Plant (CT)	Connecticut Yankee Atomic Power Company	575	PWR	West. S&W	12/62	12/62	9/63 7/66	5/64 6/67	7/24/67	8/7/67	12/67	1/68

	Plant	Company	MWe	Type	NSSS/A-E								
17	San Onofre Nuclear Gen. Station, Unit 1 (CA)	Southern Calif. Edison & San Diego Gas & Elec. Co.	430	PWR	West. Bech.	4/60	1/63	2/63 11/65	3/64 3/67	6/14/67	7/16/67	9/67	1/68
18	N. Reactor/WPPSS (WA)h	Washington Public Power Supply System and AEC	850	GR	Burns & Roe	4/62	4/63	NA	NA	12/31/63	4/8/66	7/66	
19	Nine Mile Point Nuclear Station, Unit 1 (NY)	Niagara Mohawk Power Corporation	625	BWR	GE O	7/63	10/63	3/64 6/67	4/65 8/69	9/5/69	11/9/69	1/70	12/69
20	Oyster Creek Nuclear Power Plant, Unit 1 (NJ)	Jersey Central Power & Light Company	640	BWR	GE B&R	5/63	12/63	3/64 1/67	12/64 4/69	5/3/69	9/23/69	12/69	12/69
21	Dresden Nuclear Power Station, Unit 2 (IL)	Commonwealth Edison Company	809	BWR	GE S&L	2/65	2/65	4/65 11/67	1/66 12/69	1/7/70	4/13/70	10/70	8/70
22	Fort St. Vrain Nuclear Generating Sta. (CO)	Public Service Company of Colorado*	330	HTGR	GAC S&L	3/65	3/65	10/66 11/69	9/68 12/73	1/31/74		/74	/74
23	R.E. Ginna Nuclear Power Plant, Unit 1 (NY)	Rochester Gas & Electric Company	490	PWR	West. Gil	8/65	8/65	11/65 1/68	4/66 9/69	11/9/69	12/2/69	3/70	7/70
24	Pilgrim Station, Unit 1 (MA)	Boston Edison Company	664	BWR	GE Bech.	8/65	8/65	6/67 1/70	8/68 6/72	6/16/72	7/19/72	10/72	12/72
25	Millstone Nuclear Power Station, Unit 1 (CT)	The Millstone Point Company	652.1	BWR	GE Ebas.	4/65	9/65	11/65 3/68	5/66 10/70	10/26/70	11/29/70	12/70	3/71
26	Indian Point Station Unit 2 (NY)	Consolidated Edison Company	873	PWR	West. UEC	11/65	11/65	12/65 10/68	10/66 10/71	5/22/73	6/26/73	12/73	8/73
27	Turkey Point Station Unit 3 (FL)	Florida Power and Light Company	693	PWR	West. Bech.	11/65	11/65	3/66 5/69	4/67 7/72	10/20/72		4/74	12/72
28	Dresden Nuclear Power Station, Unit 3 (IL)	Commonwealth Edison Company	809	BWR	GE S&L	1/66	1/66	2/66 11/67	10/66 1/71	1/31/71	7/22/71	10/71	10/71
29	Palisades Plant (MI)	Consumers Power Company of Michigan	700	PWR	Comb. Bech.	1/66	1/66	6/66 11/68	3/67 3/71	5/24/71	12/31/71		12/71
30	H.B. Robinson S.E. Plant, Unit 2 (SC)	Carolina Power and Light Company	700	PWR	West. Ebas.	1/66	1/66	7/66 11/68	4/67 8/70	9/20/70	9/26/70	2/71	3/71
31	Point Beach Nuclear Plant, Unit 1 (WI)	Wisconsin Elec Pwr Co and Wisc-Mich Pwr Co	497	PWR	West. Bech.	2/66	2/66	8/66 3/69	7/67 10/70	11/2/70	11/6/70	4/71	12/70
32	Monticello Nuclear Generating Plant (MN)	Northern States Power Company	545	BWR	GE Bech.	4/66	4/66	8/66 11/68	6/67 9/70	12/10/70	3/5/71	6/71	6/71
33	Quad-Cities Station, Unit 1 (IL)	Com. Edison-Iowa-Illinois Gas & Electric	800	BWR	GE S&L	4/66	4/66	5/66 9/68	2/67 9/71	10/25/71	4/12/72	12/72	8/72
34	Browns Ferry Nuclear Pwr Plant, Unit 1 (AL)	Tennessee Valley Authority	1065	BWR	GE O	6/66	6/66	7/66 9/70	5/67 6/73	8/17/73	10/15/73	3/74	6/74
35	Browns Ferry Nuclear Pwr Plant, Unit 2 (AL)	Tennessee Valley Authority	1065	BWR	GE O	6/66	6/66	7/66 9/70	5/67	2Q/74		3Q/74	10/74

Table B-1 continued

Project/Location	Owner	Cap. Net (MWe)	Type	NSSS/ AE Contr.	Public Annc'd.	NSSS Contr. Award	CP/ OL Applied	CP/ OL Issued	Initial Crit.	First Elec.	Initial Design Power	Commercial Oper.
36 Oconee Nuclear Station, Unit 2 (SC)	Duke Power Company	886	PWR	B&W O/Bech.	7/66	7/66	11/66 6/69	11/67 2/73	4/19/73	5/6/73	11/73	7/73
37 Oconee Nuclear Station, Unit 2 (SC)	Duke Power Company	886	PWR	B&W O/Bech.	7/66	7/66	11/66 6/69	11/67 10/73	11/11/73	12/5/73	3/74	1974
38 Quad Cities Station, Unit 2 (IL)	Com. Edison-Iowa-Illinois Gas & Electric	800	BWR	GE S&L	7/66	7/66	8/66 9/68	2/67 3/72	4/26/72	5/23/72	12/72	10/72
39 Peach Bottom Atomic Pwr Station, Unit 2 (PA)	Philadelphia Electric Co. PSE&G, ACEC, DP&LC	1065	BWR	GE Bech.	5/66	8/66	2/67 8/70	1/68 8/73	9/16/73		5/74	6/74
40 Peach Bottom Atomic Pwr Station, Unit 3 (PA)	Philadelphia Electric Co. PSE&G, ACEC, DP&LC	1065	BWR	GE Bech.	5/66	8/66	2/67 8/70	1/68	7/74		10/74	12/74
41 Salem Nuclear Generating Station, Unit 1 (NJ)	Public Service Elec & Gas Co, PEC, ACEC, DP&LC	1090	PWR	West. O	5/66	8/66	12/66 8/71	9/68	5/75		6/75	9/75
42 Vermont Yankee Generating Station (VT)	Vermont Yankee Nuclear Power Corporation	513.9	BWR	GE Ebas.	12/65	8/66	11/66 1/70	12/67 3/72	3/24/72	9/20/72	11/72	11/72
43 Fort Calhoun Station, Unit 1 (NB)	Omaha Public Power District	457.4	PWR	Comb. GHDR	6/66	10/66	4/67 11/69	6/68 5/73	8/5/73	8/25/73	5/74	9/73
44 Surry Power Station, Unit 1 (VA)	Virginia Electric & Power Company	788	PWR	West. S&W	6/66	10/66	3/67 1/70	6/68 5/72	7/1/72	7/4/72	11/72	12/72
45 Surry Power Station, Unit 2 (VA)	Virginia Electric & Power Company	788	PWR	West. S&W	10/66	10/66	3/67 1/70	6/68 1/73	3/7/73	3/10/73	4/73	5/73
46 Diablo Canyon Nuc Power Plant, Unit 1 (CA)	Pacific Gas & Electric Company	1084	PWR	West. O	9/66	11/66	1/67 10/73	4/68	2/75		2/75	9/75
47 Three Mile Island Nuclear Station, Unit 1 (PA)	Metropolitan Edison Co.	819	PWR	B&W Gil	11/66	11/66	5/67 3/70	5/68 4/74	5/74		7/74	10/74
48 Bailly Generating Station (IN)	Northern Indiana Public Service Company	660	BWR	GE S&L	1/67	1/67	8/70	5/74	1979		1979	1979
49 Crystal River Plant, Unit 3 (FL)	Florida Power Corp.	825	PWR	B&W Gil	2/67	2/67	8/67 2/71	9/68	3/75		3/75	3/75
50 Kewaunee Nuclear Power Plant, Unit 1 (WI)	Wisconsin Group (WPSC, WP&LC, MG&EC)	541	PWR	West. PSE	2/67	2/67	8/67 1/71	8/68 12/73	3/7/74		4/74	5/75
51 Maine Yankee Atomic Power Plant (ME)	Maine Yankee Atomic Power Corporation	790	PWR	Comb. S&W	1/66	2/67	9/67 8/70	10/68 9/72	10/23/72	11/8/72	12/72	12/72
52 Point Beach Nuclear Plant, Unit 2 (WI)	Wisconsin Elec Pwr Co & Wisc.-Mich. Power Co.	497	PWR	West. Bech.	2/67	2/67	7/67 3/69	7/68 11/71	5/30/72	8/2/72	4/73	10/72

No.	Plant	Utility	MW	Type	Vendor	AE								
53	Prairie Island Nuclear Gen. Plant, Unit 1 (MN)	Northern States Power Company	530	PWR	West.	PSE	2/67	2/67	3/67, 2/71	6/68, 8/73	11/73		4/74	12/73
54	Shoreham Nuclear Power Station (NY)	Long Island Lighting Co.	819	BWR	GE	S&W	4/65	2/67	5/68	4/73	2/78		3/78	5/78
55	Three Mile Island Nuclear Station, Unit 2 (PA)[i]	Jersey Central Power and Light Company	905	PWR	B&W	B&R	2/67	2/67	4/68, 4/74	11/69	5/76		7/76	5/77
56	Zion Station, Unit 1 (IL)	Commonwealth Edison Company	1050	PWR	West.	S&L	2/67	2/67	7/67, 11/70	12/68, 4/73	6/19/73	6/28/73	11/73	7/73
57	Arkansas Nuclear One, Unit 1 (AR)	Arkansas Power and Light Company	850	PWR	B&W	Bech.	4/67	4/67	11/67, 4/71	12/68, 5/74	6/74		9/74	9/74
58	Cooper Nuclear Station (NB)	Nebraska Public Power District	778	BWR	GE	B&R	6/66	4/67	7/67, 3/71	6/68, 1/74	2/21/74		4/74	5/74
59	Indian Point Station, Unit 3 (NY)	Consolidated Edison Company	965	PWR	West.	UEC	4/67	4/67	4/67, 12/70	8/69	3/75		5/75	6/75
60	Turkey Point Station, Unit 4 (FL)	Florida Power and Light Company	693	PWR	West.	Bech.	11/65	4/67	3/66, 5/69	4/57, 4/73	6/11/73	6/21/73	4/74	9/73
61	Calvert Cliffs Nuclear Power Plant, Unit 1 (MD)	Baltimore Gas and Electric Company	845	PWR	Comb.	Bech.	5/67	5/67	1/68, 1/71	7/69	7/74		9/74	10/74
62	Calvert Cliffs Nuclear Power Plant, Unit 2 (MD)	Baltimore Gas and Electric Company	845	PWR	Comb.	Bech.	5/67	5/67	1/68, 1/71	7/59	6/75		8/75	9/75
63	Oconee Nuclear Station, Unit 3 (SC)	Duke Power Company	886	PWR	B&W	O/Bech.	5/67	5/67	4/67, 6/69	11/67	6/74		9/74	6/74
64	Salem Nuclear Generating Station, Unit 2 (NJ)	Public Service Elec & Gas Co., PEC, ACEC, DP&LC	1115	PWR	West.	O	5/66	5/67	10/67, 8/71	9/68	5/76		6/76	9/76
65	Browns Ferry Nuclear Pwr Plant, Unit 3 (AL)	Tennessee Valley Authority	1065	BWR	GE	O	6/67	6/67	7/67, 9/70	7/68	2Q/75		3Q/75	9/75
66	Prairie Island Nuclear Gen. Plant, Unit 2 (MN)	Northern States Power Company	530	PWR	West.	PSE	6/67	5/67	3/67, 2/71	6/68	10/74		11/74	11/74
67	Donald C. Cook Plant, Unit 1 (MI)	Indiana & Michigan Electric Co.	1060	PWR	West.	AEP	12/66	7/67	12/67, 2/71	3/69	9/74		2/75	4/75
68	Donald C. Cook Plant, Unit 2 (MI)	Indiana & Michigan Electric Co.	1060	PWR	West.	AEP	7/67	7/67	12/67, 2/71	3/69	12/75		3/76	5/76
69	Zion Station, Unit 2 (IL)	Commonwealth Edison Company	1050	PWR	West.	S&L	7/67	7/67	8/67, 11/70	12/68, 11/73	12/24/73		4/74	/74
70	Rancho Seco Nuclear Gen. Station (CA)	Sacramento Municipal Utility District	304	PWR	B&W	Bech.	4/66	3/67	11/67, 4/71	10/68	6/74		9/74	10/74

Table B-1 continued

Project/Location	Owner	Cap. Net (MWe)	Type	NSSS/ AE Contr.	Public Annc'd.	NSSS Contr. Award	CP/ OL Applied	CP/ OL Issued	Initial Crit.	First Elec.	Initial Design Power	Com- mercial Oper.
71 Beaver Valley Power Station, Unit 1 (PA)	Duquesne Light, Ohio Ed, and Penn. Power	852	PWR	West. S&W	9/67	9/67	1/69 10/72	6/70	1/75		3/75	5/75
72 Limerick Generating Station, Unit 1 (PA)	Philadelphia Electric Company	1065	BWR	GE Bech.	10/67	10/67	3/70		7/79		8/79	10/79
73 Limerick Generating Station, Unit 2 (PA)	Philadelphia Electric Company	1065	BWR	GE Bech.	10/67	10/67	3/70		12/80		1/81	3/81
74 North Anna Power Station, Unit 1 (VA)	Virginia Electric & Power Company	898	PWR	West. S&W	10/67	10/67	3/69 5/73	2/71	2/76		4/76	5/76
75 Edwin I. Hatch Nuclear Plant, Unit 1 (GA)	Georgia Power Company	786	BWR	GE SSC /Bech.	6/67	12/67	5/68 3/71	9/69	6/74		9/74	10/74
76 St. Lucie, Unit 1 (FL)	Florida Power and Light Company	801	PWR	Comb. E'bas.	2/68	12/67	1/69 3/73	7/70	9/75		11/75	12/75
77 Millstone Nuclear Power Station, Unit 2 (CT)	The Millstone Point Company	828	PWR	Comb. O/Bech.	6/67	12/67	2/69 8/72	12/70	2/75		4/75	5/75
78 Brunswick Steam Electric Plant, Unit 1 (NC)	Carolina Power and Light Company	821	BWR	GE UEC	1/68	1/68	7/68 10/72	2/70	6/75		11/75	12/75
79 Brunswick Steam Electric Plant, Unit 2 (NC)	Carolina Power and Light Company	821	BWR	GE UEC	1/68	1/68	7/68 10/72	2/70	8/74		11/74	1/75
80 Duane Arnold Energy Ctr., Unit 1 (IA)	Iowa Elec L&PC, Cent IPC, and Corn Belt PC	569	BWR	GE Bech.	2/68	2/68	11/68 5/72	6/70 2/74	3/23/74		1974	1974
81 Sequoyah Nuclear Power Plant, Unit 1 (TN)	Tennessee Valley Authority	1140	PWR	West. O	4/68	4/68	10/68 1/74	5/70	4Q/75		2Q/76	6/76
82 Sequoyah Nuclear Power Plant, Unit 2 (TN)	Tennessee Valley Authority	1140	PWR	West. O	4/68	4/68	10/68 1/74	5/70	3Q/76		1Q/77	2/77
83 Midland Nuclear Power Plant, Unit 1 (MI)	Consumers Power Company of Michigan[j]	492	PWR	B&W Bech.	12/67	5/68	1/69	12/72	12/78		2/80	3/80
84 Midland Nuclear Power Plant, Unit 2 (MI)	Consumers Power Company of Michigan[j]	818	PWR	B&W Bech.	12/67	5/68	1/69	12/72	12/79		2/79	3/79
85 Susquehanna Steam Elec. Station, Unit 1 (PA)	Pennsylvania Power and Light Company	1050	BWR	GE Bech.	5/67	5/68	3/71	11/73	1979		1979	1979
86 Susquehanna Steam Electric Station, Unit 2 (PA)	Pennsylvania Power and Light Company	1050	BWR	GE Bech.	5/68	5/68	3/71	11/73	1981		1981	1981

No.	Plant	Utility	MW	Type	Mfr.							
87	Diablo Canyon Nuc Power Plant, Unit 2 (CA)	Pacific Gas & Electric Company	1106	PWR	West. O	2/68	7/68	6/68 10/73	12/70	12/75	2/76	6/76
88	Enrico Fermi Atomic Power Plant, Unit 2 (MI)	Detroit Edison Company	1093	BWR	GE O/S&L	7/68	8/68	4/69	9/72	8/76	2/77	4/77
89	Davis-Besse Nuclear Power Station, Unit 1 (OH)	Toledo Ed & Cleveland Elec. Illuminating Co.	906	PWR	B&W Bech.	2/68	10/68	7/69 3/73	3/71	9/75	1/76	2/76
90	Trojan Nuclear Plant (OR)	Portland General Electric EW&EB and PP&LC	1130	PWR	West. Bech.	2/67	11/68	6/69 2/73	2/71	2/75	5/75	7/75
91	James A. FitzPatrick Nuc. Power Plant (NY)k	Power Authority of the State of New York	821	BWR	GE S&W	8/68	12/68	12/68 6/71	5/70	6/74	9/74	9/74
92	Joseph M. Farley Nuclear Plant, Unit 1 (AL)	Alabama Power Company	829	PWR	West. SSC/Bech.	5/69	5/69	10/69 8/73	8/72	8/75	11/75	12/75
93	Hope Creek Nuc Gen Station, Unit 1 (NJ)	Public Service Electric and Gas Co.	1067	BWR	GE O	8/69	8/69	2/70		12/80	4/81	5/81
94	Hope Creek Nuc Gen Station, Unit 2 (NJ)	Public Service Electric and Gas Co.	1067	BWR	GE O	8/69	8/69	2/70		12/81	4/82	5/82
95	Wm. H. Zimmer Nuc Power Station, Unit (OH)	Cincinnati Gas & Elec Co., C&SOEC&DAPL	810	BWR	GE S&L	3/68	9/69	4/70	10/72	4/77	7/77	8/77
96	William B. McGuire Nuclear Station, Unit 1 (NC)	Duke Power Company	1180	PWR	West. O	11/69	11/69	9/70 5/74	2/73	8/76	10/76	11/76
97	William B. McGuire Nuclear Station, Unit 2 (NC)	Duke Power Company	1180	PWR	West. O	11/69	11/69	9/70 5/74	2/73	6/77	8/77	9/77
98	Forked River Nuclear Gen. Station, Unit 1 (NJ)	Jersey Central Power and Light Co.	1070	PWR	Comb. B&R	12/68	12/69	6/70	7/73	11/78	4/79	5/79
99	North Anna Power Station, Unit 2 (VA)	Virginia Electric & Power Company	898	PWR	West. S&W	10/67	1/70	3/69 5/73	2/71	2/76	10/76	11/76
100	San Onofre Nuclear Gen Station, Unit 2 (CA)l	Southern Calif Edison and San Diego Gas & Elec Co	1140	PWR	Comb. Bech.	1/70	1/70	5/70	10/73	1979	1979	1979
101	San Onofre Nuclear Gen Station, Unit 3 (CA)l	Southern Calif Edison and San Diego Gas & Elec Co	1140	PWR	Comb. Bech.	1/70	1/70	5/70	10/73	1980	1980	1980
102	Edwin I Hatch Nuclear Plant, Unit 2 (GA)	Georgia Power Company	795	BWR	GE SSC /Bech.	2/70	2/70	7/70	12/72	1/78	3/78	4/78
103	Aguirre Nuclear Power Plant (PR)m	Puerto Rico Water Resources Authority	583	PWR	West. GHDR	5/70	5/70	11/70		6/79	9/79	10/79
104	Arkansas Nuclear One, Unit 2 (AR)	Arkansas Power and Light Company	912	PWR	Comb. Bech.	5/70	5/70	9/70 4/74	12/72	8/76	2/77	2/77
105	LaSalle County Nuclear Station, Unit 1 (IL)	Commonwealth Edison Co.	1078	BWR	GE S&L	3/70	5/70	11/70	9/73	8/78	11/78	12/78

Table B-1 continued

Project/Location	Owner	Cap. Net (MWe)	Type	NSSS/AE Contr.	Public Annc'd.	NSSS Contr. Award	CP/OL Applied	CP/OL Issued	Initial Crit.	First Elec.	Initial Design Power	Commercial Oper.
106 LaSalle County Nuclear Station, Unit 2 (IL)	Commonwealth Edison Co.	1078	BWR	GE S&L	3/70	5/70	11/70	9/73	1/79		4/79	5/79
107 Bellefonte Nuclear Plant, Unit 1 (AL)	Tennessee Valley Authority	1213	PWR	B&W O	8/70	8/70	6/73		2Q/79		4Q/79	12/79
108 Bellefonte Nuclear Plant, Unit 2 (AL)	Tennessee Valley Authority	1213	PWR	B&W O	8/70	8/70	6/73		1Q/80		3Q/80	9/80
109 Watts Bar Nuclear Plant, Unit 1 (TN)	Tennessee Valley Authority	1169	PWR	West. O	8/70	8/70	5/71	1/73	1Q/78		2Q/78	6/78
110 Watts Bar Nuclear Plant, Unit 2 (TN)	Tennessee Valley Authority	1169	PWR	West.	8/70	8/70	5/71	1/73	4Q/78		1Q/79	3/79
111 Waterford Generating Station, Unit 3 (LA)	Louisiana Power and Light Company	1113	PWR	Comb. Ebas.	9/70	9/70	12/70		12/78		2/79	6/79
112 Jos. M. Farley Nuclear Plant, Unit 2 (AL)	Alabama Power Company	829	PWR	West. SSC/Bech.	6/70	12/70	6/70 8/73	8/72	10/76		12/76	1/77
113 Unit 1 (CA)	Pacific Gas & Electric Co.	1128	BWR	GE Bech.	2/71	2/71			1981		1981	1981
114 Unit 2 (CA)	Pacific Gas & Electric Co.	1128	BWR	GE Bech.	2/71	2/71			1982		1982	1982
115 Virgil C. Summer Nuclear Sta. Unit 1 (SC)	S.C. Elec. & Gas Co. & S.C. Public Service Auth.	900	PWR	West. Gil	2/71	2/71	6/71	3/73	10/77		1/78	1/78
116 WPPSS Nuclear Project No. 2 (WA)	Washington Public Power Supply System	1103	BWR	GE B&R	2/67	3/71	8/71	3/73	5/77		7/77	9/77
117 Shearon Harris Station, Unit 1 (NC)	Carolina Power & Light Co.	915	PWR	West. Ebas.	4/71	4/71	9/71		4/79		8/79	10/79
118 Shearon Harris Station, Unit 2 (NC)	Carolina Power & Light Co.	915	PWR	West. Ebas.	4/71	4/71	9/71		1980		1980	10/80
119 Shearon Harris Station, Unit 3 (NC)	Carolina Power & Light Co.	915	PWR	West. Ebas.	4/71	4/71	9/71		1981		1981	10/81
120 Shearon Harris Station, Unit 4 (NC)	Carolina Power & Light Co.	915	PWR	West. Ebas.	4/71	4/71	9/71		1981		1982	3/82
121 Byron Station, Unit 1 (IL)	Commonwealth Edison Co.	1120	PWR	West. S&L	4/71	4/71	9/73		1/80		4/80	5/80

No.	Plant	Utility	MWe	Type	Vendor						
122	Byron Station, Unit 2 (IL)	Commonwealth Edison Co.	1120	PWR	West. S&L	4/71	4/71	9/73		1/81	4/81	5/81
123	North Anna Power Station, Unit 3 (VA)	Virginia Electric & Power Co.	907	PWR	B&W S&W	4/71	4/71	9/71		10/77	2/78	3/78
124	North Anna Power Station, Unit 4 (VA)	Virginia Electric & Power Co.	907	PWR	B&W S&W	4/71	4/71	9/71		11/78	2/79	3/79
125	Fulton Generating Station Unit 1 (PA)	Philadelphia Electric Company	1140	HTGR	GAC S&W	8/71	8/71	11/73		7/80	5/81	5/81
126	Fulton Generating Station Unit 2 (PA)	Philadelphia Electric Company	1140	HTGR	GAC S&W	8/71	8/71	11/73		7/82	5/83	5/83
127	Alvin W. Vogtle Nuclear Plant, Unit 1 (GA)	Georgia Power Co.	1113	PWR	SSC/Bech.	9/71	9/71	2/73		1980	1980	4/80
128	Alvin W. Vogtle Nuclear Plant, Unit 2 (GA)	Georgia Power Co.	1113	PWR	West. SSC/Bech.	9/71	9/71	2/73		1981	1981	4/81
129	Beaver Valley Power Station, Unit 2 (PA)	Duquesne Light, Ohio Ed., and Penn. Power	852	PWR	West. S&W	9/71	9/71	11/72	5/74	2/79	4/79	6/79
130	Nine Mile Point Nuclear Station, Unit 2 (NY)	Niagara Mohawk Power Corporation	1080	BWR	GE S&W	6/71	9/71	6/72		11/78	3/79	5/79
131	Summit Power Station, Unit 1 (DE)	Delmarva Power & Light Co.	770	HTGR	GAC UEC	12/71	12/71	8/73		3/80	7/80	10/80
132	Summit Power Station, Unit 2 (DE)	Delmarva Power & Light Co.	770	HTGR	GAC UEC	12/71	12/71	8/73		3/82	7/82	10/82
133	Enrico Fermi Power Plant, Unit 3 (MI)	Detroit Edison Co.	1171	BWR	GE D/S&L	1/72	1/72			1981	1981	8/81
134	Grand Gulf Nuclear Station, Unit 1 (MS)	Mississippi Power & Light Co.	1250	BWR	GE Bech.	8/71	1/72	11/72		3/79	7/79	9/79
135	Grand Gulf Nuclear Station, Unit 2 (MS)	Mississippi Power & Light Co.	1250	BWR	GE Bech.	8/71	1/72	11/72		3/81	7/81	9/81
136	Pilgrim Station, Unit 2 (MA)[n]	Boston Edison Co.	1180	PWR	Comb. Bech.	3/72	3/72	12/73		5/80	7/80	8/80
137	Greenwood Energy Center, Unit 2 (MI)	Detroit Edison Co.	1200	PWR	B&W Bech.	4/72	4/72	9/73		2/80	5/80	8/80
138	Greenwood Energy Center, Unit 3 (MI)	Detroit Edison Co.	1200	PWR	B&W Bech.	4/72	4/72	9/73		2/82	5/82	8/82
139	Vidal Nuclear Generating Station, Unit 1 (CA)	Southern California Edison Co.[o]	770	HTGR	GAC Bech.	5/72	5/72			1981	6/81	6/82
140	Vidal Nuclear Generating Station, Unit 2 (CA)	Southern California Edison Co.	770	HTGR	GAC Bech.	5/72	5/72			1982	6/82	6/83

Table B-1 continued

Project/Location	Owner	Cap. Net (MWe)	Type	NSSS/ AE Contr.	Public Annc'd.	NSSS Contr. Award	CP/ OL Applied	CP/ OL Issued	Initial Crit.	First Elec.	Initial Design Power	Com- mercial Oper.
141 Perry Nuclear Power Plant, Unit 1 (OH)	Cleveland Elec. Ill. Co. (DUQL,OHED,PEPL,TOED)	1205	BWR	GE Gil	10/71	6/72	7/73		1979		1979	1979
142 Perry Nuclear Power Plant, Unit 2 (OH)	Cleveland Elec. Ill. Co. (DUQL,OHED,PEPL,TOED)	1205	BWR	GE Gil	10/71	6/72	7/73		1980		1980	1980
143 Seabrook Nuclear Station, Unit 1 (NH)	Public Serv Co of NH & United Illuminating Co.	1200	PWR	West. UEC	3/67	6/72	7/73		7/79		10/79	11/79
144 Seabrook Nuclear Station, Unit 2 (NH)	Public Service of NH & United Illuminating Co.	1200	PWR	West. UEC	2/72	6/72	7/73		1980		1980	1980
145 River Bend Station, Unit 1 (LA)	Gulf States Utilities Co.	934	BWR	GE S&W	11/71	6/72	9/73		3/80		8/80	9/80
146 Catawba Nuclear Station, Unit 1 (SC)	Duke Power Co.	1153	PWR	West. O	7/72	7/72	10/72		11/78		1/79	3/79
147 Catawba Nuclear Station, Unit 2 (SC)	Duke Power Co.	1153	PWR	West. O	7/72	7/72	10/72		11/79		1/80	3/80
148 Atlantic Generating Station, Unit 1 (NJ)	Public Service Electric & Gas Co.;ACEC; JCPL	1150	PWR	West. OPS	5/71	9/72	3/74		12/79		4/80	5/80
149 Atlantic Generating Station, Unit 2 (NJ)	Public Service Electric & Gas Co. ACEC; JCPL	1150	PWR	West. OPS	5/71	9/72	3/74		7/80		12/80	1/81
150 Douglas Point, Unit 1 (MD)	Potomac Electric Power Co.	1178	BWR	GE Ebas.	9/72	9/72	7/73		12/79		1/80	3/80
151 Douglas Point, Unit 2 (MD)	Potomac Electric Power Co.	1178	BWR	GE Ebas.	9/72	9/72	7/73		1981		11/82	3/82
152 Braidwood Station, Unit 1 (IL)	Commonwealth Edison Co.	1120	PWR	West. S&L	9/72	9/72	9/73		1/80		4/80	5/80
153 Braidwood Station, Unit 2 (IL)	Commonwealth Edison Co.	1120	PWR	West. S&L	9/72	9/72	9/73		6/81		9/81	10/81
154 Surry Power Station, Unit 3 (VA)	Virginia Electric & Power Co.	859	PWR	B&W S&W	9/72	9/72	5/73		11/79		2/80	3/80
155 Surry Power Station, Unit 4 (VA)	Virginia Electric & Power Co.	859	PWR	B&W S&W	9/72	9/72	5/73		11/80		2/81	3/81
156 Comanche Peak Steam Electric Station, Unit 1 (TX)	Texas P & L, Dallas P&L, Texas Elec Serv	1150	PWR	West. G&H	7/72	10/72	7/73		6/79		8/79	1/80

No. / Name	Utility	MW	Type	Vendor						
157 Comanche Peak Steam Electric Station, Unit 2 (TX)	Texas P & L, Dallas P&L, Texas Elec Serv	1150	PWR	West. G&H	7/72	10/72	7/73	6/79	8/79	1/82
158 Clinch River Breeder Reactor Plant (TN)	U.S. Government	350	LMFB	West. B&R	8/72	11/72		3/79	12/79	1980
159 St. Lucie, Unit 2 (FL)	Florida Power & Light Co.	801	PWR	Comb. Ebas.	11/72	11/72	5/73	9/80	11/80	12/80
160 WPPSS Nuclear Project, Unit 1 (WA)p	Washington Public Power Supply System	1206	PWR	B&W UEC	11/72	11/72	10/73	4/80	7/80	9/80
161 Quanicasse, Unit 1 (MI)	Consumers Power Co.	1150	PWR	West. Bech.	12/72	12/72	2/74	1981		1981
162 Quanicasse, Unit 2 (MI)	Consumers Power Co.	1150	PWR	West. Bech.	12/72	12/72	2/74	1982		1982
163 TVA Plant 1, Unit 1	Tennessee Valley Authority	1205	BWR	GE	5/72	12/72		2Q/80	4Q/80	12/80
164 TVA Plant 1, Unit 2	Tennessee Valley Authority	1205	BWR	GE	5/72	12/72		2Q/81	4Q/81	12/81
165 TVA Plant 2, Unit 1	Tennessee Valley Authority	1205	BWR	GE	5/72	12/72		4Q/80	2Q/81	6/81
166 TVA Plant 2, Unit 2	Tennessee Valley Authority	1205	BWR	GE	5/72	12/72		4Q/81	2Q/82	6/82
167 Central Alabama, Unit 1 (AL)	Alabama Power Co.	1200	BWR	GE SSC/Bech.	4/72	12/72		3/81	7/81	9/81
168 Central Alabama, Unit 2 (AL)	Alabama Power Co.	1200	BWR	GE SSC/Bech.	4/72	12/72		8/82	7/82	9/82
169 Clinton Nuclear Power Plant, Unit 1 (IL)	Illinois Power Co.	955	BWR	GE S&L	2/72	1/73	10/73	12/79	6/80	6/80
170 Clinton Nuclear Power Plant, Unit 2 (IL)	Illinois Power Co.	955	BWR	GE S&L	2/72	1/73	10/73	12/82	6/83	6/83
171 Blue Hills Station, Unit 1 (TX)	Gulf States Utilities Co.	918	PWR	Comb. Bech.	11/72	2/73		8/81	8/81	9/81
172 Millstone Nuclear Power Station, Unit 3 (CT)	Northeast Utilities	1156	PWR	West. S&W	1/72	2/73	2/73	2/79	4/79	5/79
173 PGE No. 2 (OR)	Portland General Electric	1260	PWR	B&W	2/73	2/73		1/80	5/80	7/80
174 Allens Creek, Unit 1 (TX)	Houston Lighting & Power Co.	1150	BWR	GE	8/72	3/73	11/73	12/79	2/80	3/80
175 Allens Creek, Unit 2 (TX)	Houston Lighting & Power Co.	1150	BWR	GE	8/72	3/73	11/73	1980	1980	1980

Table B-1 continued

Project/Location	Owner	Cap. Net (MWe)	Type	NSSS/ AE Contr.	Public Annc'd.	NSSS Contr. Award	CP/ OL Applied	CP/ OL Issued	Initial Crit.	First Elec.	Initial Design Power	Com-mercial Oper.
176 Perkins Nuclear Station, Unit 1 (NC)	Duke Power Company	1280	PWR	Comb.	4/73	4/73	5/74		9/80		1/81	1/81
177 Perkins Nuclear Station, Unit 2 (NC)	Duke Power Company	1280	PWR	Comb.	4/73	4/73	5/74		9/80			1/82
178 Perkins Nuclear Station, Unit 3 (NC)	Duke Power Company	1280	PWR	Comb.	4/73	4/73	5/74		9/80			1/83
179 Cherokee Nuclear Station, Unit 1 (SC)	Duke Power Company	1280	PWR	Comb.	4/73	4/73	5/74		6/82		9/82	9/82
180 Cherokee Nuclear Station, Unit 2 (SC)	Duke Power Company	1280	PWR	Comb.	4/73	4/73	5/74		5/82			9/83
181 Cherokee Nuclear Station, Unit 3 (SC)	Duke Power Company	1280	PWR	Comb.	4/73	4/73	5/74		5/82			9/84
182 Jamesport, Unit 1 (NY)	Long Island Lighting Co.	1150	PWR	West. S&W	4/73	6/73			1/81			5/81
183 (WI)	Wisconsin Elec. Power; WIPS;WIPL,MAGE	900	PWR	West. S&W	7/73	7/73			1980	1980	1980	1980
184 (WI)	Wisconsin Elec. Power; WIPS;WIPL,MAGE	900	PWR	West. S&W	7/73	7/73			1982		1982	1982
185 South Texas Nuclear Project, Unit 1 (TX)	Houston Lighting & Power; CPL	1250	PWR	West. BRRT	6/73	7/73			12/80		2/81	3/81
186 South Texas Nuclear Project, Unit 2 (TX)	Houston Lighting & Power, CPL	1250	PWR	West. BRRT	6/73	7/73			1981		1981	1981
187 WPPSS Nuclear Project, Unit 3 (WA)	Washington Public Pwr Supply System	1242	PWR	Comb. Ebas.	1/73	7/73	3/74		4/81		7/81	9/81
188 Alvin W. Vogtle Nuclear Plant, Unit 3 (GA)	Georgia Power Co.	1113	PWR	West.	9/71	7/73	2/73		1/82		3/82	4/82
189 Alvin W. Vogtle Nuclear Plant, Unit 4 (GA)	Georgia Power Co.	1113	PWR	West.	9/71	7/73	2/73		1/83		3/83	4/83
190 River Bend Station, Unit 2 (LA)	Gulf States Utilities Co.	934	BWR	GE S&W	9/73	9/73	9/73		3/82		8/82	9/82
191 Palo Verde Nuclear Gen. Station, Unit 1 (AZ)	Arizona Public Service; TG&E;STRP;PSNM;EPEC	1270	PWR	Comb. Bech.	8/73	10/73			1981		1981	1981

No.	Plant	Company	MW	Type	Vendor					
192	Palo Verde Nuclear Gen. Station, Unit 2 (AZ)	Arizona Public Service; TG&E;STRP;PSNM;EPEC	1270	PWR	Comb. Bech.	8/73	10/73	1982	1982	1982
193	Palo Verde Nuclear Gen. Station Unit 3 (AZ)	Arizona Public Service; TG&E;STRP;PSNM;EPEC	1270	PWR	Comb. Bech.	8/73	10/73	1984	1984	1984
194	Atlantic Generating Stat., Unit 3	Public Service Electric & Gas	1150	PWR	West. OPS	11/73	11/73	1985	1985	1985
195	Atlantic Generating Stat., Unit 4	Public Service Electric & Gas	1150	PWR	West. OPS	11/73	11/73	1986	1986	1986
196	Black Fox No. 1 (OK)	Public Service of Oklahoma	950	BWR	GE B&V	1/73	12/73	1982		1982
197	Black Fox No. 2 (OK)	Public Service of Oklahoma	950	BWR	GE B&V	1/73	12/73	1983		1984
198	SKAGIT County (WA)	Puget Sound Power & Light Co.	1200	BWR	GE Bech.	1/73	12/73	1982	1982	1981
199	Calloway, Unit 1 (MO)	Union Electric Co.	1150	PWR	West. Bech.	12/71	7/73	1980	1980	1981
200	Wolf Creek (KS)	Kansas Gas & Electric, Kansas City P & L	1150	PWR	West. Bech.	2/73	7/73	1981		4/81
201	Tyrone Energy Park, Unit 1 (WI)	Northern States Power Co.	1150	PWR	West. Bech.	3/73	7/73	1982		4/82
202	Tyrone Energy Park, Unit 2 (WI)	Northern States Power Co.	1150	PWR	West. Bech.	3/73	7/73	1983		10/83
203	Calloway, Unit 2 (MO)	Union Electric Co.	1150	PWR	West. Bech.	7/73	7/73	1982		4/83
204	ROGE/SNUPPS (NY)	Rochester Gas & Electric Co.	1150	PWR	West. Bech.	7/73	7/73	1982		10/82
205	Davis-Besse Nuclear Power Station, Unit 2 (OH)	Toledo Edison Co; OHED; CLEI; DUQL, PEPL	905	PWR	B&W	8/73	12/73	1981	1981	6/81
206	Davis-Besse Nuclear Power Station, Unit 3 (OH)	Toledo Edison Co.; OHED; CLEI; DUQL, PEPL	905	PWR	B&W	8/73	12/73	1982	1982	1/83
207	Wm. T. Zimmer Nuc Power Station, Unit 2 (OH)	Cincinnati Gas & Elec Co., C&SOEC&DAPL	1170	BWR	GE	10/73	1/74	1982		1982
208	Central Alabama, Unit 3 (AL)	Alabama Power Co.	1200	BWR	GE SSC/Bech.	1/74	1/74	3/83	7/83	9/83
209	Central Alabama, Unit 4 (AL)	Alabama Power Co.	1200	BWR	GE SSC/Bech.	1/74	1/74	3/84	7/84	9/84
210	Jamesport Unit 2 (NY)	Long Island Lighting Co.	1150	PWR	West. S&W	2/74	2/74	1/83		5/83

Table B-1 continued

Project/Location	Owner	Cap. Net (MWe)	Type	NSSS/AE Contr.	Public Annc'd.	NSSS Contr. Award	CP/OL Applied	CP/OL Issued	Initial Crit.	First Elec.	Initial Design Power	Commercial Oper.
211 Florida Power Unit 2 (FL)	Florida Power Corp.	1300	PWR	Comb.	3/74	3/74						1983
212 Florida Power Unit 3 (FL)	Florida Power Corp.	1300	PWR	Comb.	3/74	3/74						1986
213 St. Rosalie Unit 1 (LA)	Louisiana Power & Light Co.	1200	HTGR	GAC UEC	3/74	3/74						1982
214 St. Rosalie Unit 2 (LA)	Louisiana Power & Light Co.	1200	HTGR	GAC UEC	3/74	3/74						1984
215 Offshore No. 1 (FL)	Jacksonville Elec. Authority	1150	PWR	West.	4/74	4/74						1982
216 Offshore No. 2 (FL)	Jacksonville Elec. Authority	1150	PWR	West.	4/74	4/74						1984
217 NEES-1 (RI)	New England Power	1200	PWR	West.	4/66	5/74			1980		1980	1981
218 NEES-2 (RI)	New England Power	1200	PWR	West.	1/71	5/74			1982		1982	1983
219 PGE No. 3	Portland General Elec.	1260	PWR	BWR	5/74	5/74						1983
220 Blue Hills Station, Unit 2 (TX)	Gulf States Utilities Co.	918	PWR	Comb. Bech.	2/73	5/74			3/83		8/83	9/83
A Virgil C. Summer Nuclear Sta., Unit 2 (SC)	South Carolina Electric & Gas Co.	900			2/71				10/77			5/78
B Lower Lehigh 1 (PA)	Pennsylvania Power & Light Co.	1100			1/72				1983		1983	1983
C Lower Lehigh 2 (PA)	Pennsylvania Power & Light Co.	1100			1/72				1985		1985	1985
D Montague Nuclear Power Station, Unit 1 (MA)	Northeast Utilities	1150			1/72							1981
E NYSEG/GE Unit 1 (NY)	New York State Electric & Gas Co.	1220		UEC	12/72				1982		1982	1982
F ERIE, Unit 1 (OH)	Ohio Edison Co.; TOED, CLEI, DUQL, PEPL	1200			8/73				1982			1982
G ERIE, Unit 2 (OH)	Ohio Edison Co.; TOED, CLEI, DUQL, PEPL	1200			8/73				1983			1983

	Unit	Owner	MWe	NSSS/AE	Date			
H	Marble Hill Unit 1 (IN)	Public Service Indiana	900		11/73	1983	1983	1983
I	NYSEG/GE Unit 2 (NY)	New York State Electric & Gas Co.	1220	UEC	11/73	1984	1984	1984
J	Pilgrim Unit 3 (MA)	Boston Edison Co.	1180	PWR	11/73	1982	1982	1982
K	Montague Unit 2 (MA)	Northeast Utilities	1150		12/73			1983
L	CMED Unit 14	Commonwealth Edison Co.	1100		1/74			1983
M	CMED Unit 15	Commonwealth Edison Co.	1100		1/74			1984
N	Marble Hill Unit 2 (IN)	Public Service Indiana	900		2/74			1984
O	(NY)	Power Authority, State of New York	1200		2/74			1982
P	San Joaquin Nuclear Project 1 (CA)	LA Dept. of Water PG&E, SCE, SDE&G, CDWR	1300		3/74			**1982**
Q	San Joaquin Nuclear Project 2 (CA)	LA Dept. of Water PG&E, SCE, SDE&G, CDWR	1300		3/74			1983
R	San Joaquin Nuclear Project 3 (CA)	LA Dept. of Water PG&E, SCE, SDE&G, CDWR	1300		3/74			1985
S	San Joaquin Nuclear Project 4 (CA)	LA Dept. of Water PG&E, SCE, SDE&G, CDWR	1300		3/74			1987
T	Ft. Calhoun Station, Unit 2 (NB)	Omaha Public Pow. District	1150		4/74			1983

Source: AEC-WASH 1208, June 1974. Information in this table is in accordance with utility reporting for the end of the first quarter 1974.

Table B-1 continued

CP Construction permit
NA Not applicable
OL Operating license

[a] Decision to decommission announced 11/29/72.
[b] Plant was shut down October 1967; on 9/6/68 NSP announced plans to install gas-fired boilers for operation summer 1969; license to possess but not operate issued May 14, 1969.
[c] Shut down 9/64.
[d] Plant shut down 2/68.
[e] Shut down January 1967.
[f] Shut down for repairs January 1966; operating contract terminated 12/67.
[g] Decision to decommission announced 6/68. Order to dismantle issued 8/11/69.
[h] AEC owns reactor. WPPSS the generating facilities with Burns & Roe the contractor for WPPSS.
[i] This unit originally planned as Oyster Creek 2; transfer to Three Mile Island Site announced by GPU 12/31/68.
[j] Consumers Midland Unit 1 will also produce 3.6 million pounds per hour of process steam; Unit 2 will provide 0.4 million pounds per hour.
[k] PASNY took over Niagara Mohawk contract for Easton Plant announced and contracted in 1966.
[l] Schedule indefinite pending resolution of seismology, geology, and environmental conditions.
[m] The PRWRA is reevaluating the project.
[n] Boston Edison has an option for a duplicate unit for operation in the early 1980s.
[o] Southern California Edison has an option for two larger HTGRs at another site.
[p] This WPPS unit will use existing turbine capacity (850 MWe Net) currently in operation using N Reactor steam.

Significant Milestones Index

103 Aguirre	9 Elk River	18 N Reactor	144 Seabrook No. 2
174 Allens Creek No. 1	F ERIE 1	217 NEES-1	81 Sequoyah No. 1
175 Allens Creek No. 2	G ERIE 2	218 NEES-2	82 Sequoyah No. 2
57 Arkansas One No. 1		19 Nine Mile No. 1	1 Shippingport
104 Arkansas One No. 2	92 Farley No. 1	130 Nine Mile No. 2	54 Shoreham
80 Arnold No. 1	112 Farley No. 2	74 North Anna No. 1	198 SKAGIT County No. 1
148 Atlantic No. 1	5 Fermi No. 1	99 North Anna No. 2	185 South Texas No. 1
149 Atlantic No. 2	88 Fermi No. 2	123 North Anna No. 3	186 South Texas No. 2
194 Atlantic No. 3	133 Fermi No. 3	124 North Anna No. 4	76 St. Lucie No. 1
195 Atlantic No. 4	91 Fitzpatrick	E NYEG/GE No. 1	159 St. Lucie No. 2
	211 Florida Power 2	I NYEG/GE No. 2	213 St. Rosalie 1
48 Bailly	212 Florida Power 3		214 St. Rosalie 2
71 Beaver Valley No. 1	98 Forked River No. 1	36 Oconee No. 1	115 Summer No. 1
129 Beaver Valley No. 2	43 Fort Calhoun No. 1	37 Oconee No. 2	A Summer No. 2

107 Belleforte No. 1
108 Belleforte No. 2
13 Big Rock Pt.
196 Black Fox 1
197 Black Fox 2
171 Blue Hills No. 1
220 Blue Hills No. 2
14 BONUS
152 Braidwood No. 1
153 Braidwood No. 2
34 Browns Ferry No. 1
35 Browns Ferry No. 2
65 Browns Ferry No. 3
78 Brunswick No. 1
79 Brunswick No. 2
121 Byron No. 1
122 Byron No. 2

199 Callaway No. 1
203 Callaway No. 2
61 Calvert Cliffs No. 1
62 Calvert Cliffs No. 2
146 Catawba No. 1
147 Catawba No. 2
167 Central Alabama No. 1
168 Central Alabama No. 2
208 Central Alabama No. 3
209 Central Alabama No. 4
179 Cherokee No. 1
180 Cherokee No. 2
181 Cherokee No. 3
158 Clinch River
169 Clinton No. 1
170 Clinton No. 2
L CMED-14
M CMED-15
156 Comanche Peak No. 1
157 Comanche Peak No. 2
67 Cook No. 1
68 Cook No. 2
58 Cooper
49 Crystal River No. 3
11 CVTR

89 Davis-Besse No. 1

T Fort Calhoun No. 2
22 Ft. St. Vrain
125 Fulton No. 1
126 Fulton No. 2

23 Ginna No. 1
134 Grand Gulf No. 1
135 Grand Gulf No. 2
137 Greenwood No. 2
138 Greenwood No. 3

16 Haddam Neck
7 Hallam
117 Harris No. 1
118 Harris No. 2
119 Harris No. 3
120 Harris No. 4
75 Hatch No. 1
102 Hatch No. 2
93 Hope Creek No. 1
94 Hope Creek No. 2
8 Humboldt No. 3

2 Indian Point No. 1
26 Indian Point No. 2
59 Indian Point No. 3

182 Jamesport No. 1
210 Jamesport No. 2

50 Kewaunee No. 1

15 Lacrosse
105 LaSalle No. 1
106 LaSalle No. 2
72 Limerick No. 1
73 Limerick No. 2
B Lower Lehigh No. 1
C Lower Lehigh No. 2

51 Maine Yankee
H Marble Hill No. 1
N Marble Hill No. 2
96 McGuire No. 1
97 McGuire No. 2

53 Oconee No. 3
215 Offshore No. 1
216 Offshore No. 2
20 Oyster Creek No. 1

29 Palisades
191 Palo Verde No. 1
192 Palo Verde No. 2
193 Palo Verde No. 3
6 Pathfinder
10 Peach Bottom No. 1
39 Peach Bottom No. 2
40 Peach Bottom No. 3
176 Perkins No. 1
177 Perkins No. 2
178 Perkins No. 3
141 Perry No. 1
142 Perry No. 2
173 PGE No. 2
219 PGE No. 3
113 PG&E/GE No. 1
114 PG&E/GE No. 2
24 Pilgrim No. 1
136 Pilgrim No. 2
J Pilgrim No. 3
12 Piqua
31 Point Beach No. 1
52 Point Beach No. 2
53 Prairie Island No. 1
66 Prairie Island No. 2
O Public Authority, NY

33 Quad Cities No. 1
38 Quad Cities No. 2
161 Quanicasse No. 1
162 Quanicasse No. 2

70 Rancho Seco
145 River Bend No. 1
190 River Bend No. 2
30 Robinson No. 2
204 ROGE/SNUPPS

41 Salem No. 1
64 Salem No. 2

131 Summit No. 1
132 Summit No. 2
44 Surry No. 1
45 Surry No. 2
154 Surry No. 3
155 Surry No. 4
85 Susquehanna No. 1
86 Susquehanna No. 2

47 Three Mile No. 1
55 Three Mile No. 2
90 Trojan No. 1
27 Turkey Point No. 3
60 Turkey Point No. 4
163 TVA/Plant 1 No. 1
164 TVA/Plant 1 No. 2
165 TVA/Plant 2 No. 1
166 TVA/Plant 2 No. 2
201 Tyrone No. 1
202 Tyrone No. 2

42 Vermont Yankee
139 Vidal Nuc. Gen. Station No. 1
140 Vidal Nuc. Gen. Station No. 2
127 Vogtle No. 1
128 Vogtle No. 2
188 Vogtle No. 3
189 Vogtle No. 4

111 Waterford No. 3
109 Watts Bar No. 1
110 Watts Bar No. 2
183 WIEP/WEST No. 1
184 WIEP/WEST No. 2
200 Wolf Creek
160 WPPSS Project No. 1
116 WPPSS Project No. 2
187 WPPSS Project No. 3

4 Yankee

95 Zimmer No. 1
207 Zimmer No. 2
56 Zion No. 1
69 Zion No. 2

Table B-1 continued

205	Davis-Besse No. 2
206	Davis-Besse No. 3
46	Diablo No. 1
87	Diablo No. 2
150	Douglas Point No. 1
151	Douglas Point No. 2
3	Dresden No. 1
21	Dresden No. 2
28	Dresden No. 3

83	Midland No. 1
84	Midland No. 2
25	Millstone No. 1
77	Millstone No. 2
172	Millstone No. 3
D	Montague No. 1
K	Montague No. 2
32	Monticello

P	San Joaquin 1
Q	San Joaquin 2
R	San Joaquin 3
S	San Joaquin 4
17	San Onofre No. 1
100	San Onofre No. 2
101	San Onofre No. 3
143	Seabrook No. 1

Table B-2
Nuclear Steam Supply System Contract Awards,[a] U.S. Central Station Reactors
(no. of units/net MWe)

Year	GE No.	GE MWe	West. No.	West. MWe	B&W No.	B&W MWe	Comb. No.	Comb. MWe	GAC No.	GAC MWe	Totals[b] No.	Totals[b] MWe
Thru 1964	5	1,605.0	5	1,287.0	1	265.0	1	16.5	1	40.0	20	4,341.3
1965	3	2,125.1	3	2,056.0	—	—	—	—	1	330.0	7	4,511.1
1966	9	7,727.9	6	4,947.0	3	2,591.0	2	1,157.4	—	—	20	16,423.3
1967	7	6,238.0	13	10,841.0	5	4,270.0	5	4,109.0	—	—	30	25,458.0
1968	7	6,225.0	4	4,516.0	3	2,216.0	—	—	—	—	14	12,957.0
1969	3	2,944.0	3	3,189.0	—	—	1	1,070.0	—	—	7	7,203.0
1970	3	2,951.0	5	4,648.0	2	2,426.0	4	4,305.0	—	—	14	14,330.0
1971	4	4,439.0	10	9,878.0	2	1,814.0	—	—	4	3,820.0	20	19,951.0
1972	14	16,591.0	13	14,196.0	5	5,324.0	2	1,981.0	2	1,540.0	36	39,632.0
1973	8	8,244.0	16	18,032.0	3	3,072.0	11	13,650.0	—	—	38	42,998.0
1974	3	3,570.0	5	5,850.0	1	1,260.0	3	3,518.0	2	2,400.0	14	16,598.0
Totals	66	62,660.0	83	79,440.0	25	23,238.0	29	29,806.9	10	8,130.0	220	204,402.7

Source: AEC-WASH 1208, June 1974.

[a]Through June 1, 1974.

[b]Includes seven units with 1,127.8 MWe capacity ordered from "Other" contractors prior to 1964.

Table B-3
U.S. Nuclear Capacity as of 1914

Status	No. Units	Capacity MWe (Net)
Decommissioned	7	261.3
Lic. to oper.[a]	46	28,058.4
Building[b]	54	51,373.0
Ordered	113	124,710.0
Announced	20	22,970.0
Total	240	227,372.7

Source: AEC-WASH 1208, June 1974.
[a]Received fuel load license; not permanently shut down.
[b]Construction permit issue.

Table B-4
Achievement of Commercial Operation,[a] Nuclear Power Reactors Ordered as of June 1, 1974

Year	NWe Net		No. of Units	
	Annual	Cumulative	Annual	Cumulative
Thru 1970[b]		6,087.3		22
1971	3,456.1	9,543.4	6	28
1972	5,545.9	15,089.3	8	36
1973	5,277.4	20,366.7	7	43
1974	14,755.0	35,121.7	18	61
1975	13,016.0	48,137.7	14	75
1976	8,303.0	56,440.7	8	83
1977	7,972.0	64,412.7	8	91
1978	5,668.0	70,080.7	6	97
1979	20,677.0	90,757.7	20	117
1980	28,069.0	118,826.7	27	144
1981	31,674.0	150,500.7	28	172
1982	26,480.0	176,980.7	24	196
1983>	27,422.0	204,402.7	24	220

Source: AEC-WASH 1208, June 1974.
[a]Based on 3/31/74 schedule information furnished by utilities.
[b]Includes 7 plants which operated in prior years but have since been permanently shut down: Hallam (75 MWe); Piqua (11.4 MWe); CVTR (17 MWe); BONUS (16.5 MWe); Pathfinder (58.5 MWe); Elk River (22 MWe); FERMI 1 (60.9 MWe).

Appendix C
Technical Notes: Basic
Nuclear Concepts

Atoms

The simplest atoms—those of hydrogen—have a nucleus of one proton plus either zero, one or two neutrons, around which a single electron revolves, much like the earth revolves around the sun. Helium atoms have a nucleus of two protons and one or more neutrons, with two planetary electrons revolving about the nucleus. Lithium atoms have three protons and two or more neutrons with three planetary electrons, and so on.

  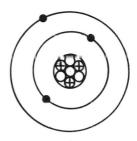

Element #1   Element #2   Element #3
Hydrogen-1   Helium-4   Lithium-7

All atoms having the same number of protons in the nucleus are collectively called an *element*. The number of protons is called the *atomic number* of the element. The total number of protons plus neutrons is called the *mass number* of that particular kind of atom of the element (see *isotopes* in following section).

In building the elements, nature has arranged for a new and distinct elementary substance for each proton added to the nucleus. But neutrons are also needed, perhaps as a sort of "nuclear glue" to keep the positively charged protons from repelling each other. There are apparently "right" numbers and "wrong" numbers of neutrons for every element. The atoms of an element having the "right" number of neutrons in relation to the number of protons are stable and unchanging with time. Those atoms having the "wrong" number of neutrons are unstable (see *radioactivity* in a following section) and spontaneously undergo change over a period of time, such as to achieve a stable configuration.

Source: AEC-WASH 1250, July, 1973.

Isotopes

There are three kinds of hydrogen atoms in nature, differing only in the number of neutrons in the nucleus:

| Hydrogen-1 | Hydrogen-2 Deuterium | Hydrogen-3 Tritium |

All behave similarly in chemical reactions (the planetary electrons govern the chemistry of an element), but are slightly different in physical properties, such as density. Hydrogen-1 is the most commonly occurring hydrogen atom, constituting 99.985% of all hydrogen atoms on earth. Hydrogen-2 (deuterium) makes up most of the remaining 0.015%, while hydrogen-3 (tritium) occurs only in trace quantities. Tritium is unstable and undergoes spontaneous change (see radioactivity).

Hydrogen-1, -2 and -3 are called isotopic forms of hydrogen or, more simply, *isotopes* of the element hydrogen. All of the elements have two or more isotopes, although some isotopes are unstable, as mentioned above.

Nuclides

In descriptions of processes that primarily involve the nuclei of the isotopes of the various elements, the nuclei are frequently called *nuclides*. The isotopies of a given element are different *nuclides*, so that there are as many nuclides as there are isotopes of elements. The system used to describe nuclides involves the symbol for the element preceded by a subscript indicating the atomic number and a superscript indicating the mass number of the isotope. Thus, the three hydrogen nuclides (isotopes) would be written:

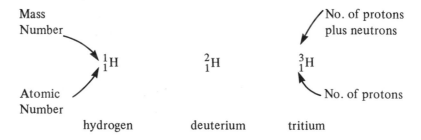

Mass Number

Atomic Number

hydrogen   deuterium   tritium

No. of protons plus neutrons

No. of protons

As another example, the element having thirty-six protons is the nuclei of its isotopes is called krypton. There are twenty-two known isotopes of krypton, six of which are stable and the remainder unstable (radioactive). All twenty-two nuclides have the same atomic number, but different mass numbers, depending on the number of neutrons in the nuclide. These range from $^{74}_{36}Kr$ to $^{95}_{36}Kr$.

Radioactivity, Radionuclides

There are sixty-four naturally occurring nuclides and a larger number of artificially produced (in nuclear reactors, particle accelerators) nuclides which are unstable and spontaneously undergo change. The vast majority of the unstable nuclides of interest in this report undergo change by one of two routes:

1. By the spontaneous emission of a beta particle (electron) *from the nucleus*, usually accompanied by gamme radiation (electromagnetic radiation like X-rays, but usually more energetic).
2. By the spontaneous emission of an alpha particle (a newly formed helium atom, stripped of its two planetary electrons, so that it bears a double positive electric charge $^{4}_{2}He^{++}$). The alpha particle is emitted *from the nucleus* and is also accompanied by gamma radiation.

The phenomenon of spontaneous nuclear disintegration is called *radioactivity*, and the disintegrating nuclide is said to be undergoing *radioactive decay*. Each radioactive species (also called *radioisotope* or *radionuclide*) decays with a characteristic and unvarying time rate, which is measured by its *half-life*.

The half-life of a radionuclide is the time for half of any given number of atoms of that nuclide to undergo radioactive decay. If one starts the clock when there are N radioactive atoms of a given kind, there will be $\frac{1}{2} N$ atoms remaining after one half-life has passed, $\frac{1}{4} N$ after two half-lives, $1/8 N$ after three half-lives, and so on. The half-lives of the various radionuclides cover a tremendous range of times, from fractions of seconds to billions of years. The shorter the half-life, the more rapid is the rate of radioactive decay.

In suitable instruments, the beta and alpha particles emitted by decaying radionuclides can be detected and counted, so that the number of atoms decaying per unit of time can be determined. The unit used to measure amounts of radioactivity is the *curie*, which is defined as that quantity of any radionuclide decaying at a rate of 3.7×10^{10} disintegrations per second. A sample of radioactive material is said to have an activity of 1 *curie* when it disintegrates at this rate.

Ions, Ionizing Radiation

Atoms, with their equal numbers of positively charged protons and negatively charged planetary electrons, are normally electrically neutral. There are many chemical and physical processes, however, wherein atoms and molecules (combinations of atoms) either lose or gain planetary electrons. Those which lose electrons bear a net positive electric charge; those that gain electrons have a net negative electric charge. Such charged atoms or molecules are called *ions*.

The alpha, beta, and gamma radiation emitted by radionuclides are called *ionizing radiation*, because they are sufficiently energetic to dislodge planetary electrons from some of the atoms or molecules they encounter while traveling through matter. The "debris" left in the path of such ionizing radiation consists of *ion pairs*—positive and negative fragments of disrupted molecules, or freed planetary electrons and the positively charged residual atom. The ions formed revert to electric neutrality fairly quickly by recapturing electrons, or by

Table C-1
Partial List of Radionuclides

Element	Nuclide	Particle Emitted	Half-Life	Number of Curies per Gram of Nuclide
Plutonium	$^{239}_{94}Pu$	alpha	2.4×10^4 Years	0.06
Uranium	$^{238}_{92}U$	alpha	4.5×10^9 Years	1.9×10^{-7}
	$^{235}_{92}U$	alpha	7.1×10^8 Years	1.2×10^{-6}
	$^{233}_{92}U$	alpha	1.6×10^5 Years	0.01
Thorium	$^{232}_{90}Th$	alpha	1.4×10^{10} Years	1.1×10^{-7}
Radium	$^{226}_{88}Ra$	alpha	1.6×10^3 Years	1.0
	$^{224}_{88}Ra$	alpha	3.64 days	1.7×10^5
Radon	$^{222}_{86}Rn$	alpha	3.8 days	1.6×10^5
Cesium	$^{137}_{55}Cs$	beta	30 years	87.3
Xenon	$^{133}_{54}Xe$	beta	5.3 days	1.9×10^5
Iodine	$^{131}_{53}I$	beta	8.06 days	1.25×10^5
Strontium	$^{90}_{38}Sr$	beta	28.9 years	138
Krypton	$^{85}_{36}Kr$	beta	10.7 years	395
Potassium	$^{40}_{19}K$	beta	1.3×10^9 years	6.9×10^{-6}
Carbon	$^{14}_{6}C$	beta	5,730 years	4.5
Tritium	$^{3}_{1}H$	beta	12.3 years	9,700

combining with nearby ions of opposite charge. By the latter process, some molecules quite different from those originally present can be formed.

Radioactive Decay Products

In radioactive decay by alpha particle emission, the immediate product of decay is a nuclide with *two protons* and *two neutrons less* than the decaying nuclide. It will, therefore, always be an isotope of the element with an atomic number two less and a mass number four less than the parent nuclide. For example:

$$^{238}_{92}U \longrightarrow ^{234}_{90}Th + ^{4}_{2}He^{++} \text{ (alpha particle)}$$

The product, in this case, is also radioactive and decays by beta emission to protoactinium-234, which in turn is radioactive and decays as shown in the following table.

$$^{234}_{90}Th \longrightarrow ^{234}_{91}Pa + \text{beta particle}$$

Note that the product of beta decay has one more proton and one less neutron than the parent and is, therefore, an isotope of the element with an atomic number one greater than the parent nuclide.

Decay through many successive daughter products is characteristic of the heaviest nuclides, those beyond lead (atomic number 82). With only a few exceptions, decay by beta emission is characteristic of the lighter radionuclides, those below lead on the atomic number scale. For these radionuclides, beta decay often produces a stable nuclide directly, as for example:

$$^{3}_{1}H \longrightarrow ^{3}_{2}He + \text{beta particle}$$
$$\text{Tritium Helium-3}$$

Nuclear Fission and Nuclear Reactors

Neutrons

In the late 1920s the emission of a very penetrating radiation was observed when the alpha radiation from polonium was made to impinge on certain light elements, such as beryllium, boron, and lithium. In 1932, this penetrating radiation was identified as a neutral particle (given the name, neutron), ejected from the nuclei of the light elements after absorption of an alpha particle. For example,

$$^{9}_{4}Be + ^{4}_{2}He^{++} \longrightarrow ^{12}_{6}C + ^{1}_{0}N$$

Table C-2
The Uranium Series

Element	Symbol	Radiation Emitted	Half-Life
Uranium	^{238}U	alpha	4.51×10^9 years
Thorium	^{234}Th	beta	24.1 days
Protactinium[a]	^{234}Pa	beta	1.18 minutes
Uranium	^{234}U	alpha	2.48×10^5 years
Thorium	^{232}Th	alpha	8.0×10^4 years
Radium	^{226}Ra	alpha	1.62×10^3 years
Radon	^{222}Rn	alpha	3.82 days
Polonium[b]	^{218}Po	alpha (99.98%) and beta (0.02%)	3.05 minutes
Lead	^{214}Pb	beta	26.8 minutes
Astatine	^{218}At	alpha	2 seconds
Bismuth[b]	^{214}Bi	beta (99.96%) and alpha (0.01%)	19.7 minutes
Polonium	^{214}Po	alpha	1.6×10^{-4} second
Thallium	^{210}Tl	beta	1.32 minutes
Lead	^{210}Pb	beta	19.4 years
Bismuth[b]	^{210}Bi	beta ($\sim$100%) and alpha (2×10^{-4}%)	5.0 days
Polonium	^{210}Po	alpha	138.4 days
Thallium	^{206}Tl	beta	4.20 minutes
Lead	^{206}Pb	Stable	

[a]Protactinium also undergoes a process of isomeric transition in 0.12% of its decay events. The resulting isomer of ^{234}Pa has a lower energy state. It then undergoes beta decay, with a half-life of 6.7 hours, to form ^{234}U.

[b]Undergoes both alpha and beta decay, in definite proportion to the decay events, as shown.

With convenient laboratory sources of neutrons, based on alpha radiation of light elements, many scientists of the time began studying the interaction of neutrons with matter.

Fission

The phenomenon of fission (splitting apart) of uranium nuclei upon absorption of a neutron was discovered in 1939. It was established soon thereafter that the uranium isotope, $^{235}_{92}U$ (frequently written: U-235) is the uranium nuclide which most readily fissions, particularly with low energy (slowly moving) neutrons.

U-235 constitutes only 0.711% of natural uranium, the most abundant isotope being U-238 at 99.284%. U-238 will undergo fission with energetic (fast moving) neutrons but not with slow neutrons. Thus, U-238 is fissionable with sufficiently energetic neutrons, but U-235 is a *fissile* nuclide—easily undergoing fission with low energy neutrons, and fissionable by neutrons of all energies.

Fission Products

When U-235 and other fissile or fissionable nuclides undergo fission, they split into two fragments, sometimes three, called *fission products*. The fission products are nuclides of lighter elements (see table C-3) which recoil from the split nucleus at high speeds for short distances. This energy of motion (kinetic

Table C-3
Principal Primary Fission Products

Isotope	Symbol	Half-Life
Strontium	^{29}Sr	53 days
Strontium	^{30}Sr	28 years
Yttrium	^{30}Y	64.2 hours
Yttrium	^{91}Y	57 days
Zirconium	^{95}Zr	65 days
Niobium	^{95}Nb	35 days
Molybdenum	^{99}Mo	68.3 hours
Ruthenium	^{103}Ru	39.8 days
Ruthenium	^{106}Ru	1 year
Rhodium	^{103m}Rh	57 minutes
Rhodium	^{106}Rh	30 seconds
Tellurium	^{132}Te	77.7 hours
Iodine	^{131}I	8.1 days
Iodine	^{132}I	2.4 hours
Xenon	^{133}Xe	5.27 days
Cesium	^{137}Cs	30 years
Barium	^{137}Ba	2.6 minutes
Barium	^{140}Ba	12.8 days
Lanthanum	^{140}La	40 hours
Cerium	^{141}Ce	32.5 days
Cerium	^{144}Ce	290 days
Praseodymium	^{143}Pr	13.7 days
Praseodymium	^{144}Pr	17.5 minutes
Neodymium	^{147}Nd	11 days
Promethium	^{147}Pm	2.6 years
Promethium	^{149}Pm	54 hours

energy) is quickly transmitted to surrounding atoms and molecules by successive collisions, thereby raising the temperature of the surroundings. The heat or thermal energy of a substance is the energy of motion—the kinetic energy—of its atoms and molecules (combinations of atoms).

Most of the fission products are radioactive and decay with their characteristic half-lives to stable nuclides, either directly or through the decay of radioactive daughter products (secondary fission products). Nearly all fission products decay by beta emission, usually accompanied by gamma radiation. Approximately 88% of the heat produced from fission comes from the kinetic energy of the fission fragments. The balance comes from the interaction of beta and gamma radiation and neutrons with surrounding atoms.

Fission Neutrons and Nuclear Chain Reactions

Neutrons are also ejected from fissioning nuclides, the average number of neutrons per fission event ranging from about 2 to about 3, depending on the fissioning nuclide and the energy of the neutron causing fission. The fission neutrons are "born" with tremendous speeds (kinetic energy), but these speeds generally drop off due to successive "billiard ball" collisions with surrounding nuclei, and many neutrons are absorbed by nearby nuclei of any of the atoms present. Nevertheless, it is possible to arrange fissile materials and structural materials in a nuclear reactor core such that a self-sustained chain reaction of fissions can be made to take place, and to stop, at will. In order for this to happen, one neutron from each fission event must manage to find another fissile or fissionable nucleus and cause it to fission, and so on.

Thermal Neutrons, Moderators

There are two ways to improve the chances for fissile nuclei to undergo fission. One is to slow down the fission neutrons with a *moderating* material, such as water. The slowing down occurs by successive collisions and is most effective with small nuclei having about the same mass as the neutron.

Water is an effective *moderator* because it contains many hydrogen-1 nuclei which are protons having almost the same mass as the neutron. When the fission neutrons have been slowed down to speeds comparable with those of the moderator molecules, they are said to be thermalized or *thermal neutrons*. That is, they have energies comparable with the thermal (heat) energy of the moderator. For a fissile nuclide, the chance of fission is increased 200-300 times with a thermal neutron, as compared with a fast (energetic) neutron.

Uranium Enrichment, Light Water Reactors

The second way to improve the chance of fission is to increase the number of fissile nuclei in the nuclear fuel. This is done by increasing the concentration of U-235 atoms in natural uranium by an isotope separation process known as the gaseious diffusion process. The fuel for most nuclear power plants in the United States is enriched in U-235 from the naturally occurring 0.72% to 3-4%. Also, these nuclear power plants predominantly use ordinary water as a moderator (a few use graphite) and as a coolant to remove the heat of fission, transform water to steam and thence to generate electricity. They are called Light Water Reactors (LWRs) to distinguish them from other power reactors (as built in Canada, for example) which use heavy water (water made with hydrogen-2, deuterium) as the moderator. In general, however, all reactors using moderators to thermalize neutrons are called thermal neutron reactors or *thermal reactors*. Those that use no moderator (or a coolant with poor moderating properties) are fast neutron or *fast reactors* (see *breeding* in a following section).

Fertile and Fissile Materials

Fission is not the only event which can occur to a fissile or fissionable nuclide upon absorption of a neutron. Actually, there are competing processes. U-235 occasionally absorbs a neutron without fissioning; it simply emits gamma radiation to rid itself of excess energy and becomes the next highest isotope or uranium, U-236, which is nonfissile.

As mentioned previously, U-238 fissions with a sufficiently energetic neutron, but it normally makes the following transformations upon absorption of neutrons of lesser energy:

$$^{238}_{92}U + \text{neutron} \longrightarrow ^{239}_{92}U \longrightarrow ^{239}_{93}Np + \text{beta} \longrightarrow ^{239}_{94}Pu + \text{beta}$$

The plutonium-239 formed by the successive beta decay of the intermediate products is fissile, like U-235, and it has a relatively long half-life (24,000 years) so that, once formed, it stays around long enough to be used as a nuclear fuel.

Thorium, $^{232}_{90}Th$, which occurs in the earth's crust with an abundance comparable to uranium, can also be the source of an artificial fissile nuclide:

$$^{232}_{90}Th + \text{neutron} \longrightarrow ^{233}_{90}Th \longrightarrow ^{233}_{91}Pa + \text{beta} \longrightarrow ^{233}_{92}U + \text{beta}$$

Fissile U-233 also has a long half-life (160,000 years) and can be used as a nuclear fuel.

In summary, there is one naturally occurring fissile nuclide, U-235, and two that can be manufactured from excess neutrons[a] in nuclear reactors, Pu-239 and U-233. The starting materials for the latter two fissile nuclides are the so-called *fertile* materials, U-238 and Th-232, respectively. All these nuclides are of great importance to the nuclear power industry.

Conversion of Fertile Material
to Fissile Material

Light water reactor fuel is generally made up of a homogeneous mixture of 3-4% U-235 atoms and 96-97% U-238 atoms, both in the form of uranium dioxide (UO_2). During operation of the reactor, some of the fission neutrons are absorbed by U-238 which converts to Pu-239. Generally, for each pound of U-235 fissioned, 0.5 to 0.6 pound of Pu-239 is formed. About half of this plutonium is fissioned in place; the balance can be recovered from the spent fuel during reprocessing. The recovered plutonium can be used in place of U-235 in some of the fuel recycled back to the reactor. Some *plutonium recycle* in light water reactors is being done by the nuclear power industry on an experimental basis at this time.

Thermal Reactor Material Balance
Per Metric Ton Uranium

U-235 Charged	34,000 grams
U-235 Discharged	9,000 grams
U-235 Consumed	25,000 grams
Fissile Pu Produced	19,000 grams
Fissile Pu Discharged	7,000 grams
Fissile Pu Consumed	12,000 grams

Th-232, U-233 (or U-235) mixtures can be used (and have been used on an experimental basis) as fuel for light water reactors. In this case, some of the thorium is converted to U-233, and the reactor is said to be operating as a *converter reactor* and the *thorium-U-233 cycle*. In the case discussed above, the reactor is a converter reactor operating on the *uranium-plutonium cycle*.

Breeding

Reactors that produce more fissile fuel (by conversion of U-238 or Th-232) than they consume in the course of producing electric power are called *breeder*

[a]That is, neutrons in excess of those needed to sustain the fission chain reaction.

reactors, and such reactors are under development for eventual commercial use on both the uranium-plutonium cycle and the thorium-U-233 cycle.

The basic differences in the design of converter reactors and breeder reactors on the *uranium-plutonium cycle* are the following:

1. The fissile fuel for breeders is Pu-239 instead of U-235. Thus, the Pu-239 produced in light water reactors can be used to start up breeder reactors.
2. The breeder core must operate on fissions caused by fast neutrons, because more neutrons are produced in fast fissions than in thermal fissions, so that more are available to convert U-238 to plutonium. This means that the coolant used to extract the heat of fission from the breeder core must be a *poor moderator*, in order that the fast fission neutrons will not be slowed down too much.
3. To compensate for the decreased chance of fission with fast neutrons, the concentration of fissile nuclei in the fuel must be increased considerably over that used in light water reactors. Breeder fuel generally needs to have a concentration of 15-20% plutonium; the remaining 80-85% can be either natural uranium or depleted uranium (the "waste" product of gaseous diffusion plants, having a lower than natural U-235 content).
4. The breeder core needs to be surrounded by a "blanket" of natural or depleted uranium to intercept neutrons that would otherwise escape from the core and be wasted.

As a consequence of the fast neutron population in the fast breeder core, about 20% of the fissions are fast neutron fissions of U-238. This is sometimes referred to as a "fast fission bonus." By comparison, about 5% of the fissions in a light water reactor are fast fissions of U-238, caused by those fission neutrons that manage to encounter a U-238 nucleus before being slowed down by the moderator.

Interestingly enough, breeding on the thorium-U-233 cycle can evidently be accomplished with thermal neutrons, because the number of neutrons emitted from the thermal fission of U-233 is sufficient for breeding if used efficiently (the neutron yield from *thermal* fission of Pu-239 is not). Also, the neutron yield from U-233 does not increase greatly as the energy of the neutron causing fission increases, whereas the neutron yield from Pu-239 fission does not become adequate for breeding until fast neutron fission predominates.

The theoretical *breeding ratio* (ratio of fissile atoms produced to those consumed) in thermal thorium-U-233 breeders is about 1.05, as compared with about 1.6 in fast, uranium-plutonium breeders (1.2 to 1.4 actually expected in practice). Accordingly, the main design problem for thorium-U-233 breeders is one of selecting and arranging core materials to minimize waste of neutrons by "parasitic" absorption in materials that do not contribute to the fission or breeding process. There are not as many excess neutrons to work with in this cycle, as in the uranium-plutonium cycle.

Fission Energy

The fission of a single U-235 nucleus (or other fissile or fissionable nucleus) is accompanied by the release of an amount of thermal energy which is about 50 million times as much as that obtained from the combustion of an atom of carbon-12, the major constituent of coal. Thus, the fissioning of one pound of fissile material produces as much heat as the burning of about 1,400 tons of coal. The heat liberated in the fissioning of all of the atoms in one pound of fissile material is approximately:

$$0.9 \times 10^{13} \text{ calories}$$
$$1.0 \times 10^{7} \text{ kwh}$$
$$3.6 \times 10^{10} \text{ Btu}$$

Bibliography

Books

Allardice, Corbin, and Trapnell, Edward R. *The Atomic Energy Commission.* New York: Praeger Publishers, 1974.

Berman, William, and Hydeman, Lee M. *Atomic Energy Commission and Regulating Nuclear Facilities.* Ann Arbor: University of Michigan Law School Atomic Energy Research Project, 1961.

Bright, James R., and Schoeman, Milton, eds. *A Guide to Practical Technological Forecasting.* Englewood Cliffs: Prentice-Hall, Inc., 1973.

Curtis, Richard, and Hogan, Elizabeth. *Perils of the Peaceful Atom.* London: Victor Gollancz, 1970.

Dawson, Frank G. *Nuclear Power: Development and Management of a Technology.* Seattle and London: University of Washington Press, 1976.

Ebbin, Steven, and Kasper, Raphael. *Citizen Groups and the Nuclear Power Controversy: Uses of Scientific and Technological Information.* Cambridge, Mass.: MIT Press, 1974.

Finkel, Arthur J., ed. *Energy, the Environment, and Human Health.* Acton, Mass.: Publishing Sciences Group, 1974.

Foreman, Harry, M.D., ed. *Nuclear Power and the Public.* Minneapolis: University of Minnesota Press, 1970.

Gofman, J.W., and Tamplin, A. *Poisoned Power: The Case Against Nuclear Power Plants.* Emmaus, Pa.: Rodale Press, 1971.

Green, Harold P., and Rosenthal, Alan, *Government of the Atom.* New York: Atherton Press, 1963.

Hewlett, Richard, and Duncan, Francis. *Nuclear Navy 1946-1962.* Chicago: University of Chicago Press, 1974.

Kasper, Raphael G., ed. *Technology Assessment: Understanding the Social Consequences of Technological Applications.* New York: Praeger Publishers, 1972.

Lewis, Richard S. *The Nuclear Power Rebellion.* New York: Viking Press, 1972.

Lowrance, William W. *Of Acceptable Risk.* Los Altos, Cal.: William Kaufman, Inc., 1976.

Murphy, Arthur W., ed. *The Nuclear Power Controversy.* Englewood Cliffs, N.J.: Prentice-Hall, Inc., 1976.

Nuclear Energy Policy Study Group. *Nuclear Power: Issues and Choices* (Sponsored by the Ford Foundation). Cambridge, Mass.: Ballinger Publishing Co., 1977.

Shapiro, Jacob. *Radiation Protection: A Guide for Scientists and Physicians.* Cambridge, Mass.: Harvard University Press, 1972.

Talbot, Allen R. *Power Along the Hudson: The Storm King Case . . . and the Birth of Environmentalism.* New York: Dutton, 1972.

Webb, Richard E. *The Accident Hazards of Nuclear Power Plants.* Amherst: University of Massachusetts Press, 1976.

Willrich, Mason, and Taylor, Theodore B. *Nuclear Theft: Risks and Safeguards.* Cambridge, Mass.: Ballinger Publishing Co., 1974.

Articles

Avins, Alfred. "Licensing Standards for Atomic Energy Licensee," *Wyoming Law Review* 11 (Fall 1956).

Brady, David, and Althoff, Phillip. "The Politics of Regulation: The Case of the Atomic Energy Commission and the Nuclear Industry," *American Politics Quarterly*, July 1973.

Bright, G.O. "Some Effects of Public Intervention on the Reactor Licensing Process," *Nuclear Safety* 13:1 (January-February 1972).

Bush, S.H., "The Role of the Advisory Committee on Reactor Safeguards in the Reactor Licensing Process," *Nuclear Safety* 13:1 (January-February 1972).

Case and Schoenbad. "Electricity or the Environment," *California Law Review*, June 1973.

Cavers, David F. "Administrative Decision-making in Nuclear Facilities Licensing," *University of Pennsylvania Law Review* 110 (January 1962).

Cornell, Nina; Noll, Roger; and Weingart, Barry. "Safety Regulation," *Setting National Priorities; the Next Ten Years.* Harry Owen, Charles L. Schultze, ed. Washington, D.C.: The Brookings Institution, 1976.

Cottrell, W.B. "ECCS Rulemaking Hearing," *Atomic Energy Law Journal* 16 (Winter 1975), 350-415.

Davis, Kenneth Kulp. "Nuclear Facilities Licensing: Another View," *University of Pennsylvania Law Review* 110:330 (1962).

Dunlayey, D.C. "Government Regulation of Atomic Industry," *University of Pennsylvania Law Review* 105 (January 1957).

Finlayson, Fred. "A View from the Outside," *Bulletin of the Atomic Scientists*, September 1975.

Gillette, Robert. "Nuclear Safety I-IV," *Science*, September 1, 8, 15, and 22, 1972.

_____. "Nuclear Safety: AEC Report Makes the Best of It," *Science*, January 21, 1973.

Gofman, John W., and Tamplin, Arthur R. "Nuclear Power, Technology and Environmental Law," *Environmental Law*, Winter 1971.

Goodman, Michael. "National Radiation Health Standards—A Study in Scientific Decisionmaking," *Atomic Energy Law Journal* 6:3 (Fall 1974).

Green, Harold P. "Nuclear Power: Risk, Liability, and Indemnity," *Michigan Law Review*, January 1973.

_____. "Nuclear Safety and the Public Interest," *Nuclear News* 15(9):75 (1972).

_____. "Safety Determinations in Nuclear Power Licensing," *Notre Dame Lawyer* 43 (June 1968).

Jacks, W. Thomas. "The Public and the Peaceful Atom: Participation in AEC Regulatory Proceedings," *Texas Law Review* 52:466 (1974).

Jopling, Gage, and Schoeman. "Forecasting Public Resistance to Technology: The Example of Nuclear Power Reactor Siting," *A Guide to Practical Technological Forecasting*. Bright and Schoeman, eds. Englewood Cliffs, N.J.: Prentice-Hall, Inc., 1973.

Keating, William Thomas. "Politics, Energy and the Environment," *American Behavioral Scientist* 19:1 (September-October 1975).

Kingsley, Sidney G. "Licensing of Nuclear Power Reactors in the U.S.," *Atomic Energy Law Journal* 7 (Winter 1965).

Kouts, Herbert J.C. "The Future of Reactor Safety Research," *Bulletin of the Atomic Scientists*, September 1975.

Minogue, Robert B. "Standards of Criteria in Use or Being Developed for Reactors," IAEA-SM-169/46, 1973.

Muntzing, L. Manning. "Standardization in Nuclear Power," *Atomic Energy Law Journal*, Spring 1973.

Niehoff, R.O. "Organization and Administration of the U.S. Atomic Energy Commission," *Public Administrative Review* 8 (September 1948).

Nuclear Safety Staff. "Regulation of Nuclear Power Reactors and Related Facilities," *Atomic Energy Law Journal* (Fall 1974).

Oulahan, G.M.C. "Atomic Energy and the Law," *Ohio Bar* 32 (June 27, 1959).

Phillips, James G. "Energy Report/Nader, Nuclear Industry Prepare to Battle over the Atom," *National Journal*, February 1, 1975.

Pike, S.T. "Model State Law for Atomic Energy Population," *Public Utilities Fortnightly* 55 (June 9, 1955).

Plaine, H.H.E. "Atomic Energy, A New Body of Administrative Law," *Journal of the Bar Association of the District of Columbia* 24 (February 1957).

Pleat and Lennemann, "Consideration for Long Term Waste, Storage and Disposal," *Atomic Energy Law Journal* 8 (September 1966).

Price, H.C. "Current Approach to Licensing Nuclear Power Plants," *Atomic Energy Law Journal* 15 (Winter 1974).

_____. "Licensing Nuclear Power Plants," *Atomic Energy Law Journal* (Winter 1972).

Primack, Joel. "Nuclear Reactor Safety: An Introduction to the Issues," *Bulletin of the Atomic Scientist*, September 1975.

Ramey, James T. "Next 5 Years in Nuclear Development and Regulation," *Atomic Energy Law Journal* 10 (Summer 1968).

_____. "Old and New Concepts in Siting and Licensing Nuclear Power Plants," *Forum* 9 (Winter 1973).

Rasmussen, Norman C. "The Safety Study and Its Feedback," *Bulletin of the Atomic Scientists*, September 1975.

Schultze, Charles. "The Public Use of the Private Interest," *Harper's*, 254:1524 (May 1977). From the Godkin lectures delivered in December 1976.

Shapar, H.K. "Licensing Nuclear Power Reactors in the U.S.," *Atomic Energy Law Journal* 15 (Fall 1973).

Tsivoglu, E.C. "Nuclear Power: The Social Conflict," *Environmental Science and Technology*, May 1971.

Upton, A.E. "Licensing and Services to Licensees and Others," *George Washington Law Review* 24 (April 1956).

Weinberg, Alvin M. "The Moral Imperatives of Nuclear Energy," *Nuclear News*, 14:12 (December 1971).

_____. "Science and Trans-Science," *Minerva* 10 (1972).

Wilson, James G. "The Politics of Regulation," *Social Responsibility and the Business Predicament*. James W. McKie, ed. Washington, D.C.: Brookings Institution, 1974.

Yellin, Joel. "The Nuclear Regulatory Commission's Reactor Safety Study," *Ball Journal of Economics*, Spring 1976.

Government Documents and Reports

Allen, Wendy. *Nuclear Reactors for Generating Electricity: U.S. Development from 1946 to 1963*. Santa Monica: The Rand Corporation, R-2116-NSF, June 1977.

Ashley, R.L., ed. *Nuclear Power Reactor Siting: Proceedings of a National Topical Meeting*, American Nuclear Society, L.A. Section, February 16-18, 1965, Division of Technical Information, CONF-650201.

Federal Radiation Council. *Background Material for the Development of Radiation Protection Standards*. Washington, D.C.: Government Printing Office, 1960.

Gandara, Arturo. *Utility Decisionmaking and the Nuclear Option*. Santa Monica: The Rand Corporation, R-2148-NSF, June 1977.

Harris, Louis, and Associates, Inc. *A Survey of Public and Leadership Attitudes toward Nuclear Power Development in the U.S.* Conducted for EBASCO Services, Inc., August 1975.

Hendrickson, McDonald and Schilling. *Review of Decision Methodologies for Evaluating Regulatory Actions Affecting Public Health and Safety*. Battelle, Pacific N.W. Labr. (BNWL-2158), December 1976.

Joint Committee on Atomic Energy. *Hearings on AEC Authorizing Legislation*. Washington, D.C.: Government Printing Office, all years, 1950-1975.

_____. *Hearings on AEC Regulatory Problems*. 87th Congress, 2nd Session. Washington, D.C.: Government Printing Office, April 1962.

_____. *Hearings on the Development, Growth, and State of the Atomic Energy Industry*. Washington, D.C.: Government Printing Office, 1955, 1956, and 1965.

_____. *Hearings on the Environmental Effects of Producing Electric Power.*

Washington, D.C.: Government Printing Office, October-November 1969 and January-February 1971.

_____. *Hearings on the Federal Radiation Council Protective Action Guides.* 89th Congress, 1st Session. Washington, D.C.: Government Printing Office, June 1965.

_____. *Hearings on Governmental Indemnity and Reactor Safety.* Washington, D.C.: Government Printing Office, May-June 1956 and March 1957.

_____. *Hearings on Licensing and Regulation of Nuclear Reactors.* 90th Congress, 1st Session. Washington, D.C.: Government Printing Office, April, May, and September 1967.

_____. *Hearings on Licensing Procedure and Related Legislation.* 92nd Congress, 1st Session. Washington, D.C.: Government Printing Office, June and July 1971.

_____. *Hearings on Nuclear Power Plant Siting and Licensing.* 93rd Congress, 2nd Session. Washington, D.C.: Government Printing Office, 1974.

_____. *Hearings on Nuclear Reactor Safety.* 93rd Congress, 1st Session. Washington, D.C.: Government Printing Office, January, September, and October 1973.

_____. *Hearings on Radiation Safety and Regulation.* 87th Congress, 1st Session. Washington, D.C.: Government Printing Office, June 1961.

_____. *Improving the AEC Regulatory Process: Committee Report.* Washington, D.C.: Government Printing Office, March 1961.

_____. *A Study of AEC Procedures and Organization in the Licensing of Nuclear Facilities: Committee Report.* Washington, D.C.: Government Printing Office, 1957.

Mooz, William E. *Cost Analysis of Light Water Reactor Plants.* Santa Monica: The Rand Corporation, R 2304 DOE, June 1978.

Perry, Robert, and others. *Development and Commercialization of the Light Water Reactor, 1946-1976.* Santa Monica: The Rand Corporation, R-2180-NSF, June 1977.

Rolph, Elizabeth S. *Regulation of Nuclear Power: The Case of the Light Water Reactor.* Santa Monica: The Rand Corporation, R-2104-NSF, June 1977.

S.M. Stoller Corp. *Central Station Nuclear Power.* March 1976, unpublished.

U.S. Atomic Energy Commission. *Draft of Reactor Safety Study: An Assessment of Accident Risks in U.S. Commercial Nuclear Plants*, Wash. 1400. Washington, D.C.: Government Printing Office, August 1974.

_____. *Nuclear Power and the Environment: Proceedings.* Washington, D.C.: Government Printing Office, September 1969.

_____. *The Safety of Nuclear Power Reactors and Related Facilities*, Wash. 1250. Washington, D.C.: Government Printing Office, July 1975.

Index

Index

Accidents, 17, 18-19; incredible, 66, 69, 85; maximum credible, 68-69, 148; worst-case, 49, 149

Advisory Committee on Reactor Safeguards (ACRS), 23, 24, 34-43 *passim*, 71, 74, 76, 85, 86-88, 89, 94, 95, 97

Advisory Committee on X-Ray and Radium Protection, 108. *See also* National Committee on Radiation Protection and Measurement

American National Standards Institute, 66, 134

American Society of Mechanical Engineers, 66, 85

Anderson, Clinton, 39, 42

Army Corps of Engineers, 107

Atomic Energy Act (1946), 21-22

Atomic Energy Act (1954), 1, 27; proposed amendment (1971), 116, 119n

Atomic Energy Commission (AEC): created (1946), 21; organization, 33-35; reorganization, 42-49, 133; abolished (1974), 116n; Division of Biology and Medicine, 34; Division of Civilian Applications, 33-34, 43 (*see also* Division of Licensing and Regulation; Office of Industrial Development); Division of Compliance, 45; Division of Inspection, 32; Division of Licensing and Regulation, 43, 45, 74, 75, 76, 77; Division of Reactor Development and Technology, 12, 23, 24, 33, 34, 35, 43, 45, 68, 70, 75, 76, 86; Division of Reactor Safety Research, 153; Division of Waste Management, 114; Hazards Evaluation Branch, 34, 35-36, 38, 39, 41, 43; Limited Work Authorization (1974), 137; Naval Reactors Branch, 24, 26, 56; Nuclear Information Center, 86;

Office of Health and Safety, 45, 50; Office of Industrial Development, 43; Office of Radiation Standards, 45; Regulatory Standards Directorate, 133, 134; Rules of Practice (1972), 137

Atomic Industrial Forum, 152

Atomic Safety and Licensing Board, 48, 71, 72-73, 74, 103, 105, 113

Babcock and Wilcox, 125

Backfitting, 58, 59, 73-74, 120n, 134

Baltimore Gas and Electric Company, 106

Beck, Clifford, 50, 61, 67

Becquerel, Antoine Henri, 11

Big Rock Point reactor, 144

Bikini nuclear tests, 108

Blowdown, defined, 19

Bodega Head Association, 64

Bodega Head reactor, 64

Boiler and Pressure Vessel Code, 66

Boiling water reactor (BWR), 18, 56

Brookhaven Labs study, 49

Brown's Ferry reactor, 79

Bupp, Irvin C., 157

Burney, Leroy E., 109

Bush, S.H., 89

California, 63

Calvert Cliffs Coordinating Committee, 106, 132, 133

Cancer, 112n

Chain reaction, 13

Closed cycle cooling, 104, 105

Code of Federal Regulations, 66

Codes, defined, 66

Comey, David, 101, 102n

Commonwealth Edison, 33, 56, 86-88

Consolidated Edison, 33, 62, 86-88

Consumer Power Company, 105

Containment failure, defined, 19

Containment structure, 60

About the Author

Elizabeth Rolph came to the subject of regulation with extensive experience both as a participant in and as an observer of government decision-making. In 1964 after completing graduate work in political science at the University of California, Berkeley, she joined the U.S. Information Services in Hong Kong and Southeast Asia. Following her tour, she returned to work in the domestic political arena. After serving several years as a science and public affairs specialist for public television in Boston and Chicago, she rejoined government as a policy analyst for the California state assembly, specializing in transportation finance and energy issues. For the past seven years Ms. Rolph has been a resident consultant at The Rand Corporation, continuing her work on urban and energy issues.

Selected Rand Books

Armor, David J.; Polich, J. Michael; and Stambul, Harriet B. *Alcoholism and Treatment*. New York: John Wiley and Sons, Inc., 1978.

Becker, Abraham S. *Military Expenditure Limitation for Arms Control: Problems and Prospects. With a Documentary History of Recent Proposals*. Cambridge, Mass.: Ballinger Publishing Company, 1977.

De Salvo, Joseph S., ed. *Perspectives on Regional Transportation Planning*. Lexington, Mass.: Lexington Books, D.C. Heath and Company, 1973.

Goldhamer, Herbert. *The Adviser*. New York: Elsevier North-Holland, Inc., 1978.

Kakalik, James S., and Wildhorn, Sorrel. *The Private Police: Security and Danger*. New York: Crane, Russak and Company, 1977.

Mitchell, Bridger; Manning, Willard G., Jr.; and Acton, Jan Paul. *Peak-Load Pricing: European Lessons for U.S. Energy Policy*. Cambridge, Mass.: Ballinger Publishing Company, 1978.

Park, Rolla Edward. *The Role of Analysis in Regulatory Decision-Making*. Lexington, Mass.: Lexington Books, D.C. Heath and Company, 1973.

Pincus, John, ed. *School Finance in Transition: The Courts and Educational Reform*. Cambridge, Mass.: Ballinger Publishing Company, 1974.

Quade, E.S. *Analysis for Public Decisions*. New York: American Elsevier Publishing Company, 1975.

Wirt, John G.; Lieberman, Arnold J.; and Levien, Roger E. *R&D Management Methods Used by Federal Agencies*. Lexington, Mass.: Lexington Books, D.C. Heath and Company, 1975.

Yin, Robert K.; Heald, Karen A.; and Vogel, Mary E. *Tinkering with the System: Technological Innovations in State and Local Services*. Lexington, Mass.: Lexington Books, D.C. Heath and Company, 1977.